Basement Physics

The Colloquial Guide

Megan Lim

Note to Reader: Physics can be a very intimidating subject. However, if you understand the fundamentals thoroughly you will find yourself able to apply even the simplest of concepts to difficult problems. The first half of this book simplifies the basic but crucial ideas of several major concepts. The second half is a collaboration of various physics problems ranging from difficult to basic. I did not write these problems myself but rather assembled a collection of problems I found interesting and deconstructed each of them. I attempted to point out many major concepts many may have missed.

Part I: Theory

1D Kinematics

kinematics - how objects move

$$V_f^2 = V_i^2 + 2a\Delta x$$

$$V_f = V_i + at$$

$$x = x_0 + V_0 t + \tfrac{1}{2}at^2$$

$$\bar{V} = \frac{V_i + V_f}{2}$$

Key: only use these equations with <u>constant</u> acceleration

* know these equations then know how to apply them

average speed = $\dfrac{\text{distance traveled}}{\text{time elapsed}}$ average velocity = $\dfrac{\text{displacement}}{\text{time elapsed}}$

instantaneous velocity $V = \lim\limits_{\Delta t \to 0} \dfrac{\Delta x}{\Delta t} = \dfrac{dx}{dt}$

average acceleration = $\dfrac{\text{change in velocity}}{\text{time elapsed}}$

instantaneous acceleration $a = \lim\limits_{\Delta t \to 0} \dfrac{\Delta v}{\Delta t} = \dfrac{dv}{dt}$

deceleration - careful: not always negative
- when magnitude of velocity is decreasing
- velocity and acceleration point in opposite directions

Derivation of Kinematic Formulas

$$a = \frac{V_f - V_i}{t} \qquad at = V_f - V_i$$

$$\boxed{V_f = at + V_i} \quad 1$$

$$\bar{V} = \frac{X_f - X_i}{t} \qquad \boxed{\bar{V} = \frac{V_i + V_f}{2}}$$

$$X_f = \bar{V}t + X_i$$

$$X_f = \left(\frac{V_i + V_f}{2}\right)t + X_i$$

$$X_f = \left(\frac{V_i + at + V_i}{2}\right)t + X_i$$

$$X_f = X_i + \left(\frac{2V_i + at}{2}\right)t \quad \rightarrow \quad \boxed{X_f = X_i + V_0 t + \tfrac{1}{2}at^2} \quad 2$$

$$\bar{V} = \frac{X_f - X_i}{t} \qquad \bar{V} = \frac{V_i + V_f}{2}$$

$$X_f = \bar{V}t + X_i$$

$$X_f = \left(\frac{V_i + V_f}{2}\right)t + X_i \qquad a = \frac{V_f - V_i}{t}$$

$$t = \frac{V_f - V_i}{a}$$

$$X_f = X_i + \left(\frac{V_i + V_f}{2}\right)\left(\frac{V_f - V_i}{a}\right)$$

$$\boxed{V_f^2 = V_i^2 + 2a\Delta x} \quad 3$$

2/3D Kinematics aka Projectile Motion

Key: use 1D kinematic formulas to break motion into x and y components then treat separately

Understand: x direction: constant velocity
: no acceleration acting on it

y direction: acceleration (gravity) always acting

TIME - the link (common component) between x and y parts

Relative Velocity

$$\vec{V}_{AC} = \vec{V}_{AB} + \vec{V}_{BC}$$

Newton & Forces

1st Law – Object continues at rest or in uniform velocity if no <u>net force</u> acts on it

2nd Law: $\boxed{F = ma}$ A net force causes an acceleration

3rd Law: <u>equal and opposite reactions</u>
When one object exerts a force on a second object, the second exerts an equal force in the opposite direction to the first

- Understand: These forces are <u>internal</u> to the system
  - aka they will cancel out

normal force: <u>perpendicular</u> to surface of contact

Key to solving problems: Σma is the sum of all the forces acting on a FBD

* it is not an actual force = never actually appears in FBD

* used for <u>calculation only</u>

Friction

kinetic friction — acts when object is actually <u>in motion</u>
— acts <u>opposite</u> to direction of object's velocity

$$\boxed{F_{fr} = \mu_K F_N}$$

*interesting: friction force is independent of area

μ_K — coefficient of kinetic friction
— depends on nature of 2 surfaces in contact

static friction — a resisting force when 2 objects in contact are <u>not</u> sliding past each other

$$\boxed{F_{fr} \leq \mu_S F_N}$$

note: unlike kinetic friction, the force is a <u>range</u>

maximum static friction = max. force needed to apply to get an object moving

Key Points: it is always either static <u>OR</u> kinetic friction present

when static friction is overcome, kinetic friction takes over

kinetic friction is always less than static friction

Circular Motion — moving at constant speed in a circle

velocity magnitude - constant
    direction - constantly changing
        - tangent to circular path

centripetal acceleration - points toward center of circle
    - always ⊥ to velocity

$$\boxed{a_R = \frac{v^2}{R}} \qquad T = \frac{1}{f} \qquad v = \frac{2\pi r}{T}$$

- frequency and period are reciprocals

$$\boxed{F = \frac{mv^2}{r}}$$

For problem solving: $\Sigma \frac{mv^2}{r}$ is used synonymously to Σma as net force acting on system

Extra: Nonuniform Circular Motion

tangential component of acceleration - acts to increase or decrease magnitude of object's velocity

$$a_{tan} = \frac{dv}{dt}$$

total acceleration vector = $\sqrt{a_{tan}^2 + a_R^2}$

Gravitation

Newton Again: Law of Universal Gravitation

$$F = G\frac{m_1 m_2}{r^2}$$

every object in the universe attracts every other particle with this force __proportional__ to product of their masses __inversely proportional__ to distance^2 between them

* a force exerted without contact

Gravity Near Earth's Surface

$$mg = G\frac{m m_E}{r_E^2}$$

subscript E = Earth property

Satellites key: maintain centripetal orbit

$$G\frac{m m_E}{r^2} = \frac{mv^2}{r}$$

Work and Energy

Work – product of displacement and the component of force parallel to the displacement

$$W = F_\parallel d = Fd\cos\theta$$

Work done by a varying force

$$W = \int_a^b \vec{F} \cdot d\vec{l}$$

* line integral

Work done by a Spring

$$W = \tfrac{1}{2}kx^2$$

Force by a Spring

$$F = -kx$$

* negative because it is a restoring force – acts in direction <u>opposite</u> to its displacement

Translational Kinetic Energy

$$K = \tfrac{1}{2}mv^2$$

Net Work – change in object's KE

$$W_{net} = \tfrac{1}{2}mv_2^2 - \tfrac{1}{2}mv_1^2$$

Understanding Work Energy Principle

$$W = \int_1^2 F \cdot dl = \int_1^2 ma\, dl$$
$$= \int_1^2 m \frac{dv}{dt} dl$$
$$= \int_1^2 m \frac{dv}{dt} \frac{dl}{dt} dt$$
$$= \int_1^2 mv\, dv$$
$$= \tfrac{1}{2}mv_2^2 - \tfrac{1}{2}mv_1^2$$

Energy Conservation

conservative force — work done by a moving object depends only on <u>initial</u> and <u>final</u> <u>position</u>

- <u>independent</u> of path taken
- net work done by an object moving in a <u>closed</u> <u>path</u> is <u>zero</u>

ex: gravity, electric fields, springs

nonconservative force — work does <u>depend on path taken</u>

ex: friction

Gravitational Potential Energy

$$U = mgh$$ - close to Earth

$$U = -\frac{GmM_E}{r}$$ - far from Earth

key: establish a reference point where $y=0$ and be consistent

General Potential Energy

$$\Delta U = -W$$

$$F(x) = \frac{-dU(x)}{dx}$$

Elastic Potential Energy

$$U = \tfrac{1}{2}kx^2$$

Energy Conservation

$$K_1 + U_1 = K_2 + U_2$$

* sum of kinetic and potential energy remains constant
* use whatever PE formula applicable to the problem

Power — rate at which work is done

$$P = \frac{W}{t}$$

$$P = \frac{F \cdot dx}{dt} = \vec{F} \cdot \vec{v}$$

Linear Momentum

$\vec{p} = m\vec{v}$ — direction of momentum is in direction of velocity

$\Sigma \vec{F} = \dfrac{d\vec{p}}{dt}$ — net external force is change in momentum

key to understanding conservation of angular momentum

* if there are no external forces acting on system, angular momentum is conserved

$$\vec{F} = \dfrac{d\vec{p}}{dt} = \dfrac{d(m\vec{v})}{dt} = m\dfrac{d\vec{v}}{dt} = m\vec{a}$$

$$\Sigma F_{ext} = 0 = \dfrac{d\vec{p}}{dt}$$ then rate of momentum change is <u>zero</u>

$$\boxed{m_A v_A + m_B v_B = m_A v_A' + m_B v_B'}$$

- as long as there is no external force acting, momentum before collision = momentum after collision

Impulse — change in momentum

$$F = \dfrac{dp}{dt} = \dfrac{J}{dt} \qquad \boxed{J = \int_{t_i}^{t_f} F\, dt}$$

Key To Solving Collision Problems

Consider Conservation of 1. Momentum
 2. Energy

Elastic Collisions - total **Kinetic Energy** is conserved

why only KE? PE only exists in that super, super short moment when 2 objects are in contact during collision
- for simplification consider only KE

1. $m_A V_A + m_B V_B = m_A V_A' + m_B V_B'$
2. $\frac{1}{2} m_A V_A^2 + \frac{1}{2} m_B V_B^2 = \frac{1}{2} m_A V_A'^2 + \frac{1}{2} m_B V_B'^2$

Inelastic Collisions - Kinetic Energy is **NOT conserved**

but momentum still is if $F_{ext} = 0$

1. $m_A V_A + m_B V_B = m_A V_A' + m_B V_B'$
2. $\frac{1}{2} m_A V_A^2 + \frac{1}{2} m_B V_B^2 = \frac{1}{2} m_A V_A'^2 + \frac{1}{2} m_B V_B'^2 +$ other forms of energy

completely inelastic - 2 objects **completely stick together**

note: ballistic pendulum problem later

Rocket Science - a variation of momentum

Key: mass decreases as lose mass of expelled gases

M (mass of rocket) dM - mass losing from gases

total system: $M + dM$

total initial momentum: $M\vec{v} + \vec{u}\,dM$

after time $t + dt$: $M + dM$
$\vec{v} + d\vec{v}$

change in momentum

                                     total system   initial momentum

$$d\vec{p} = (M + dM)(\vec{v} + d\vec{v}) - (M\vec{v} + \vec{u}\,dM)$$

$$= \cancel{M\vec{v}} + M\,d\vec{v} + \vec{v}\,dM + dM\,d\vec{v} - \cancel{M\vec{v}} - \vec{u}\,dM$$

$$= M\,d\vec{v} + \vec{v}\,dM + dM\,d\vec{v} - \vec{u}\,dM$$

                                                                    ↳ 2 differentials ≈ 0

$$d\vec{p} = M\,d\vec{v} + \vec{v}\,dM - \vec{u}\,dM$$

$$\vec{F}_{ext} = \frac{d\vec{p}}{dt} = \frac{M\,d\vec{v} + \vec{v}\,dM - \vec{u}\,dM}{dt}$$

$$\frac{d\vec{p}}{dt} = M\frac{d\vec{v}}{dt} - (\vec{u} - \vec{v})\frac{dM}{dt}$$

                                                      ↳ rel velocity to each other

↑ rocket   ↑ thrust

Mg ↓   ↓ gases

$$\boxed{\;\frac{d\vec{p}}{dt} = M\frac{d\vec{v}}{dt} - \vec{V}_{rel}\frac{dM}{dt}\;}$$

             ↓   ↓   ↓

  (F_{ext}) external force   Mass acceleration of rocket   Thrust

$$d\vec{v} = \frac{F_{ext}}{M}dt + V_{rel}\frac{dM}{M}$$

$$\int_{V_0}^{V} d\vec{v} = \int_0^t \frac{F_{ext}}{M}dt + \int_{M_0}^{M} V_{rel}\frac{dM}{M}$$

$$\boxed{\;V(t) = V_0 - gt + V_{rel}\ln\frac{M}{M_0}\;}$$

Center of Mass

weighted average of the masses relative to each of their distances to a common reference point

$$\boxed{X_{CM} = \frac{m_A X_A + m_B X_B}{m_A + m_B}}$$

note: important in dealing with problems that deal with rotation about its center of mass

$$X_{CM} = \frac{1}{M}\int x \, dm \qquad Y_{CM} = \frac{1}{M}\int y \, dm \qquad Z_{CM} = \frac{1}{M}\int z \, dm$$

$$dm = \rho \, dV$$

Rotational Motion

note: compare to translational motion

$$\theta = \frac{l}{R}$$

l - arc length
R - distance from axis of rotation

$\theta \sim x$
$\omega \sim \vec{v}$
$\alpha \sim \vec{a}$

$\omega = \frac{d\theta}{dt}$
$\alpha = \frac{d\omega}{dt}$

$l = R d\theta$

$v = \frac{dl}{dt} = R\frac{d\theta}{dt}$

$a_{tan} = \frac{dv}{dt} = R\frac{d\omega}{dt}$

$a_{rad} = \frac{v^2}{R} = \frac{(R\omega)^2}{R}$

$\omega = \omega_0 + \alpha t$
$\theta = \omega_0 t + \frac{1}{2}\alpha t^2$
$\omega^2 = \omega_0^2 + 2\alpha\theta$
$\bar{\omega} = \frac{\omega + \omega_0}{2}$

$$\omega = \frac{v}{R}$$

$$a_{tan} = R\alpha$$

$$a_{rad} = \omega^2 R$$

$a = \sqrt{a_{tan}^2 + a_{rad}^2}$

total linear acceleration

helpful formulas already covered

$f = \frac{\omega}{2\pi}$ $T = \frac{1}{f}$

Right Hand Rule: curl hand along direction of rotation, thumb points in direction of $\vec{\omega}$

$\vec{\alpha}$ - points along axis of rotation
  <u>positive</u> - points in same direction as $\vec{\omega}$
  <u>negative</u> - points in direction opposite as $\vec{\omega}$

Note: no part of object actually moves in direction of $\vec{\omega}$ and $\vec{\alpha}$

Torque - the dynamics of rotational motion

$$\tau = RF_\perp = RF\sin\theta$$

Note: $\perp$ component of force applied creates torque
$\parallel$ component of force applied creates NO torque

$$\tau = I\alpha$$

torque is the rotational analog of $F=ma$

Moment of Inertia - resistance to rotate

$$I = mR^2$$

* if there are multiple masses/point particles, sum up each I to get total moment of inertia for entire system

R - each object's distance from axis of rotation

Moment of Inertia for different shapes can be found in book

OR derive $$I = \int R^2 dm \qquad dm = \rho \, dV$$

Parallel Axis Thm

$$I = I_{cm} + Mh^2$$

when rotating axis is parallel to axis that passes through center of mass

h = distance between 2 axes

note: useful for solving problems where rotational axis is not passing through center of mass

Rotational Kinetic Energy

$$KE = \tfrac{1}{2} I \omega^2$$

$$W = \int_{\theta_1}^{\theta_2} \tau \, d\theta = \int_{\theta_1}^{\theta_2} I \omega \, d\omega = \tfrac{1}{2} I \omega_2^2 - \tfrac{1}{2} I \omega_1^2$$

Rolling Without Slipping - almost every physics rotational motion problem

* depends on the presence of static friction
* both translational and rotational kinetic energy

$$V = R\omega$$

$$K_{tot} = \tfrac{1}{2} I_p \omega^2$$

$$I_p = I_{cm} + MR^2$$

$$K_{tot} = \tfrac{1}{2} I_{cm} \omega^2 + \tfrac{1}{2} MR^2 \omega^2$$

$$K_{tot} = \tfrac{1}{2} I_{cm} \omega^2 + \tfrac{1}{2} M v_{cm}^2$$

Note: ω is same at CM and pt P, but V is not

Angular Momentum
— The rotational analog of linear momentum

$$\boxed{L = I\omega} \qquad \boxed{\Sigma \tau = \frac{dL}{dt}} \qquad \begin{aligned} \Sigma \tau &= I\alpha \\ &= I\left(\frac{d\omega}{dt}\right) \\ &= \frac{d(I\omega)}{dt} \\ \Sigma \tau &= \frac{dL}{dt} \end{aligned}$$

Conservation of Angular Momentum

if there is <u>zero net external torque</u>, angular momentum is conserved

$$\Sigma \tau = 0 = \frac{dL}{dt}$$

$$\boxed{\vec{L} = \vec{r} \times \vec{p}} \qquad \text{angular momentum expressed as a cross product}$$

$$\frac{dL}{dt} = \frac{d}{dt}(\vec{r} \times \vec{p})$$

$$\frac{dL}{dt} = \frac{d\vec{r}}{dt} \times \vec{p} + \vec{r} \times \frac{d\vec{p}}{dt} \qquad\qquad \frac{d\vec{r}}{dt} \times \vec{p} = 0$$

$$\frac{dL}{dt} = \vec{r} \times \frac{d\vec{p}}{dt} = \vec{r} \times \vec{F} = \vec{\tau}$$

Statics   aka equilibrium of a system

$\Sigma F = 0$ sum of forces acting on system is <u>zero</u>

$\Sigma \tau = 0$ sum of torques acting on system is <u>zero</u>

Fluids

weight of an object

$$\rho = \frac{m}{V}$$
$$m = \rho V$$

$$\boxed{mg = \rho V g}$$

$$\boxed{P = P_0 + \rho g h}$$

P_0 - accounts for external atmospheric pressure above liquid's surface

pressure in fluids

$$\boxed{P = \frac{F}{A}}$$

$$P = \frac{F}{A} = \frac{\rho A h g}{A} = \rho g h$$

$$\boxed{P = \rho g h}$$

pressure due to fluid at depth h is due to weight of column of liquid above it

note: area doesn't influence pressure

Buoyant Force - <u>upward</u> force acting against <u>bottom</u> of submerged object is <u>greater</u> than <u>downward</u> force acting against <u>top</u> surface of object

reason: pressure in a fluid increases with depth

Archimedes' Principle - buoyant force is equal to weight of fluid displaced by that object

basically: volume of fluid displaced = volume of submerged object

Floating Object

$$\boxed{F_B = mg}$$

buoyant force = weight of object

Fluids in Motion

mass flow rate $= \dfrac{\Delta m}{\Delta t} = \dfrac{\rho \Delta V}{\Delta t} = \dfrac{\rho A \Delta l}{\Delta t} = \rho A v$

equation of continuity: $\rho_1 A_1 v_1 = \rho_2 A_2 v_2$

$$\boxed{A_1 v_1 = A_2 v_2}$$ $\rho_1 = \rho_2$ for incompressible fluids

note: inverse relationship between area and velocity
large cross-sectional area = small velocity

Bernoulli's Equation - The Energy of Fluids

note: **inverse** relationship between fluid velocity and pressure
greater velocity = lower pressure

$$\boxed{P_1 + \tfrac{1}{2}\rho v_1^2 + \rho g y_1 = P_2 + \tfrac{1}{2}\rho v_2^2 + \rho g y_2}$$

where 1 & 2 are any 2 points along tube of flow
y is height of center of tube above a fixed level

derivation: $W_1 = F_1 \Delta l_1 = P_1 A_1 \Delta l_1$ $W_2 = -F_2 \Delta l_2 = -P_2 A_2 \Delta l_2$

$W_3 =$ work done by gravity
$W_3 = -mg(y_2 - y_1)$

$\Delta W = W_1 + W_2 + W_3$

$\tfrac{1}{2} m v_2^2 - \tfrac{1}{2} m v_1^2 = P_1 A_1 \Delta l_1 - P_2 A_2 \Delta l_2 - m g y_2 + m g y_1$

substitute $m = \rho A_1 \Delta l_1 = \rho A_2 \Delta l_2$
divide through by $A_1 \Delta l_1 = A_2 \Delta l_2$

→ get equation

Oscillations

Cause of an oscillation — restoring force is present that returns mass back to equilibrium position

Spring System — most classic example

$F = -kx$ is the restoring force that causes oscillation of mass attached to spring

Simple Harmonic Motion — explained in terms of differential equation

differential equation — a function expressed as sum of linear combinations of its ordered derivatives

Why necessary? — we guess this function so that it works

$$F = ma \qquad F = -kx$$

$$ma = -kx$$

$$ma + kx = 0$$

$$m\frac{d^2x}{dt^2} + kx = 0$$

$$\boxed{\frac{d^2x}{dt^2} + \frac{k}{m}x = 0}$$

now we guess: what x (function) will make this equation work

$$\boxed{x = A\cos(\omega t + \phi)}$$

guess is right? → let's test it

$$\frac{dx}{dt} = -\omega A \sin(\omega t + \phi)$$

$$\frac{d^2x}{dt^2} = -\omega^2 A \cos(\omega t + \phi)$$

plug back in general formula

after plugging back in

$$\left(\frac{k}{m} - \omega^2\right) A \cos(\omega t + \phi) = 0$$

works when $\boxed{\omega^2 = \frac{k}{m}}$ → $\boxed{f = \frac{1}{2\pi}\sqrt{\frac{k}{m}}}$ $\quad f = \frac{1}{T}$

max velocity of oscillator

$$v = \frac{dx}{dt} = -\omega A \sin(\omega t + \phi)$$

$$\boxed{v_{max} = \omega A}$$

max acceleration of oscillator

$$a = \frac{d^2 x}{dt^2} = -\omega^2 A \cos(\omega t + \phi)$$

$$\boxed{a_{max} = \omega^2 A}$$

Energy of a Simple Harmonic Oscillator

$$\boxed{E = \tfrac{1}{2}mv^2 + \tfrac{1}{2}kx^2 = \tfrac{1}{2}mv_{max}^2 = \tfrac{1}{2}kA^2}$$

sum of KE of object and PE of spring = maximum KE = maximum spring PE

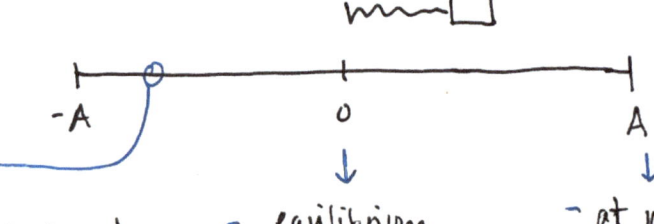

-A →
- partially displaced from equilibrium
- energy is partially KE and partially spring PE
- max acceleration

0 →
- equilibrium position
- object is at max. speed
- ALL energy is in form of $\tfrac{1}{2}mv_{max}^2$
- no acceleration because no net force of spring acting on mass

A →
- at max. displacement (amplitude)
- energy is stored ALL in form of $\tfrac{1}{2}kA^2$
- max. acceleration

The Simple Pendulum

Is it a simple harmonic oscillator?
- Is there a restoring force proportional to its displacement?

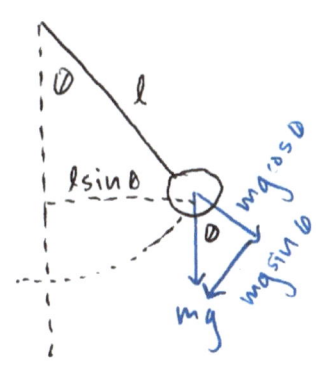

displacement of pendulum $\boxed{x = \ell\theta}$

restoring force - component of weight
 - tangent to arc $\boxed{F = -mg\sin\theta}$

$F = -mg\sin\theta$ small θ approximation
$F = -mg\theta$ $\sin\theta \sim \theta$
$F = -\frac{mg}{\ell} x$ similar to $F = -kx$?

✓ is a SHO   yay!

$\boxed{k = \frac{mg}{\ell}}$

$\omega = \sqrt{\frac{k}{m}} = \sqrt{\frac{g}{\ell}}$

$\boxed{\begin{array}{c} f = \frac{1}{2\pi}\sqrt{\frac{g}{\ell}} \\ T = 2\pi\sqrt{\frac{\ell}{g}} \end{array}}$

Damped Harmonic Motion

key: external force is acting to reduce amplitude until oscillators stop

$$F_{damping} = -bv$$

b — damping constant

$$ma = -kx - bv$$
$$ma + kx + bv = 0$$
$$m\frac{d^2x}{dt^2} + b\frac{dx}{dt} + kx = 0$$

express as differential equation

Solution: find $x(t)$ that fits differential equation

$$\boxed{x(t) = Ae^{(-\frac{b}{2m})t} \cos(w't)}$$
$$f' = \frac{1}{2\pi}\sqrt{\frac{k}{m} - \frac{b^2}{4m^2}}$$
$$w' = \sqrt{\frac{k}{m} - \frac{b^2}{4m^2}}$$

key: $w' \neq w$

- overdamped: $b^2 > 4mk$
- underdamped: $b^2 < 4mk$
- critical damping: $b^2 = 4mk$

Physical Pendulum

— an extended object that oscillates

Note: unlike simple pendulum, mass is not concentrated at end

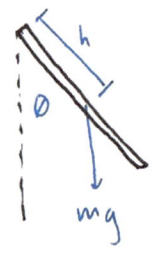

$$\tau = -mgh \sin\theta \qquad \Sigma \tau = I\alpha = I\frac{d^2\theta}{dt^2}$$

$$I\frac{d^2\theta}{dt^2} = -mgh\sin\theta$$

$$\frac{d^2\theta}{dt^2} + \frac{mgh}{I}\sin\theta = 0$$

small angle approximation $\sin\theta \sim \theta$

$$\boxed{\frac{d^2\theta}{dt^2} + \frac{mgh}{I}\theta = 0 \\ w = \sqrt{\frac{mgh}{I}}}$$

Waves
— the transport of energy from one location to another

source of a wave: oscillation

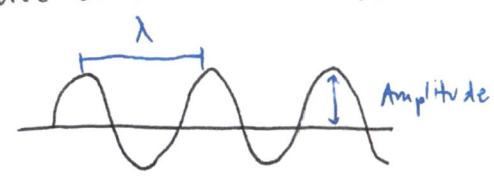

$v = \lambda f$

- each particle oscillates about an equilibrium point

$v = \dfrac{\lambda}{T}$

velocity of transverse waves

$v = \sqrt{\dfrac{F_T}{\mu}}$

μ — linear mass density

* greater F_T → greater velocity
* greater μ → lower velocity

Note: velocity of each individual particle and velocity of the wave differ in **magnitude** AND **direction**

wave: moves along direction of travel (right or left)
particle: oscillates up and down about equilibrium position

Mathematical Representation of a Traveling Wave

moving to the right

$$D(x) = A\sin\dfrac{2\pi}{\lambda}x = A\sin\left[\dfrac{2\pi}{\lambda}(x - vt)\right] = A\sin\left(\dfrac{2\pi x}{\lambda} - \dfrac{2\pi t}{T}\right)$$

$$D(x,t) = A\sin(kx - \omega t) \qquad k = \dfrac{2\pi}{\lambda} \qquad v = \dfrac{\omega}{k}$$

moving to the left

$$D(x,t) = A\sin(kx + \omega t)$$

*note the sign change when direction switches

General Form: $D(x,t) = A\sin(kx - \omega t + \phi)$

ϕ = phase angle
= adjusts for shifting of wave position to left or right depending on __initial conditions__

Standing Waves

- both ends of string must be **fixed**
- a transmitted wave is inverted when it reaches one end and is reflected backwards

Key: 2 traveling transmitted and reflected waves interact to produce a **standing wave** (does not appear to be traveling)

Energy is NOT transmitted down string but rather stands in place.

node - point of **destructive** interference
antinode - point of **constructive** interference

Mathematical Representation of a Standing Wave

$$D_1(x,t) = A\sin(kx - \omega t)$$ right traveling wave

$$D_2(x,t) = A\sin(kx + \omega t)$$ left traveling wave

$$D = D_1 + D_2 = A[\sin(kx-\omega t) + \sin(kx+\omega t)]$$

using trig identity $\sin\theta_1 + \sin\theta_2 = 2\sin\frac{1}{2}(\theta_1+\theta_2)\cos\frac{1}{2}(\theta_1-\theta_2)$

$$\boxed{D = 2A\sin kx \cos \omega t}$$ l = length of string

nodes occur at every half λ

$$l = \frac{n\lambda_n}{2} \rightarrow \lambda_n = \frac{2l}{n}$$

Last Minute Tips

- ALWAYS draw the free body diagram <u>first</u>
- You will find several problems that can be either solved using kinematics or energy
    - When choosing methods, understand <u>energy's limitations</u>:
        1. time is not easily extractable from this approach (unlike when using kinematic equations)
        2. you do not know vector components of velocity or θ between them (only find the magnitude)
- BEFORE solving the problem and making assumptions, check <u>required conditions are satisfied</u>
    - ex: for momentum conservation, check $\Sigma F_{ext} = 0$
        - ask what kind of collision is occurring?
            - inelastic or elastic → affects KE conservation
    - for angular momentum conservation, <u>check $\Sigma \tau = 0$</u>
    - for rotational kinematics, check <u>rolling without slipping</u>
- for derivations, write down all the equations you know first then try to find common linkages of variables between them

Part II: Problems

Derivation of Range (Projectile Motion) at <u>Equal Height</u>

$V_f^2 = V_i^2 + 2a\Delta x$

$V_f = V_i + at$

$x = x_o + V_o t + \frac{1}{2}at^2$

y displacement is 0

* common component between x & y components is <u>TIME</u>

y: $\quad 0 = V_o \sin\theta \, t - \frac{g}{2} t^2 \quad \rightarrow \quad 0 = t(V_o \sin\theta - \frac{g}{2} t)$

x: $\quad R = V_o \cos\theta \, t \qquad\qquad\qquad\qquad t = \dfrac{2V_o \sin\theta}{g}$

$R = V_o \cos\theta \left(\dfrac{2V_o \sin\theta}{g} \right)$

$$\boxed{R = \dfrac{\sin 2\theta \, V_o^2}{g}}$$

Trig: $2\sin\theta\cos\theta = \sin 2\theta$

* complementary angles give same range
* max range at 45°

$\sin 2\theta = 1$

$2\theta = 90$

$\theta = 45°$ max

2 Boxes Connected by a Massless String / Frictionless

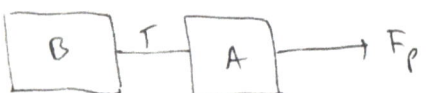

F_p = force applied

① separate

*tension at the end if the same rope is the same

KEY: set net force = ma always

1. $T = m_B a_B$
2. $F_p - T = m_A a_A$

add equations

$F_p = m_B a_B + m_A a_A$

$$a = \frac{F_p}{m_B + m_A}$$

* acceleration of boxes are the same
* Tension between A and B are the same

② consider as system

$(m_B + m_A) a_{Total} = F_p$

$$\frac{F_p}{m_B + m_A} = a_{total}$$

- F_p - net force acting on entire system
* F_p - acts ONLY on block A

Atwood's Machine (Simple Pulley)

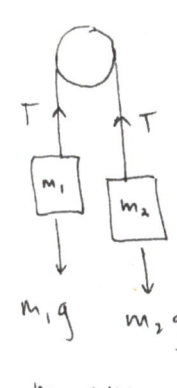

$m_1 g \quad m_2 g$

$m_1 < m_2$

KEY: identify net direction of motion in each mass

★ Tension is equal in both sides of the pulley

★ m_1 and m_2 rise and fall with same acceleration

↑ pos
↓ neg

① **Individual**

$T - m_1 g = m_1 a$ bc $m_1 < m_2$, net movement for m_1 is ↑
$m_2 g - T = m_2 a$ net movement for m_2 is ↓

add equations $m_2 g - m_1 g = m_1 a + m_2 a$

$$\boxed{a = \frac{m_2 g - m_1 g}{m_1 + m_2}}$$

② **Consider as System**

System Movement   Net Sum of individual components acting

$(m_1 + m_2) a_{tot} = m_2 g - m_1 g$

$$\boxed{a_{tot} = \frac{m_2 g - m_1 g}{m_1 + m_2}}$$

↑T   Σ   ↑T   Σ
[m_1] ↑$m_1 a$ [m_2] ↓$m_2 a$
↓   ↓
$m_1 g$ $m_2 g$

Pendulum inside Accelerating Car

- car accelerates at constant acceleration → pendulum swings

$y:\quad T\cos\theta = mg$

$x:\quad T\sin\theta = ma$

$$\frac{T\sin\theta}{T\cos\theta} = \frac{ma}{mg}$$

$$\tan\theta = \frac{a}{g} \;\rightarrow\; \boxed{\tan^{-1}\left(\frac{a}{g}\right) = \theta}$$

2 Boxes and a Pulley with Friction

* Tension throughout string is same ⎫ be connected
* m_1 and m_2 move with same acceleration ⎭ by pulley

KEY: tendency of direction for entire system to move is positive

① Individual

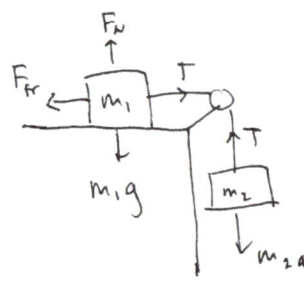

$T - \mu m_1 g = m_1 a$

$m_2 g - T = m_2 a$

_____

$m_2 g - \mu m_1 g = m_1 a + m_2 a$

$$\boxed{a = \frac{m_2 g - \mu m_1 g}{m_1 + m_2}}$$

② Consider system

$$(m_1 + m_2) a_{tot} = m_2 g - \underset{F_{fr}}{\mu m_1 g}$$

$$\boxed{a_{tot} = \frac{m_2 g - \mu m_1 g}{m_1 + m_2}}$$

4 blocks linked together

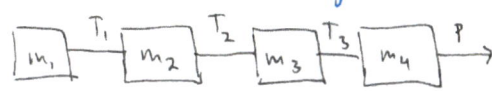

Key: BC blocks (system) move together, they share the same acceleration

$T_1 \neq T_2 \neq T_3 \neq P$

① Individual

$N = mg$ for every block bc NO vertical movement

$T_1 - \mu m_1 g = m_1 a$

$T_2 - T_1 = m_2 a$

$T_3 - T_2 = m_3 a$

$P - T_3 = m_4 a$

$P - \mu m_1 g = (m_1 + m_2 + m_3 + m_4) a$

$$a = \frac{P - \mu m_1 g}{m_1 + m_2 + m_3 + m_4}$$

② System Considered – Ignore Tensions between blocks → Cancel out

$(m_1 + m_2 + m_3 + m_4) a_{tot} = P - \mu m_1 g$

$$a_{tot} = \frac{P - \mu m_1 g}{m_1 + m_2 + m_3 + m_4}$$

Trick for Relative Velocities

$$V_{ac} = V_{ab} + V_{bc}$$

velocity of a w/ respect to c = velocity a respect to b + velocity b respect to c

★ KEY: $V_{ab} = -V_{ba}$ $V_{bc} = -V_{cb}$

→ switch signs according to convenience

Atwood Machine in Accelerating Elevator

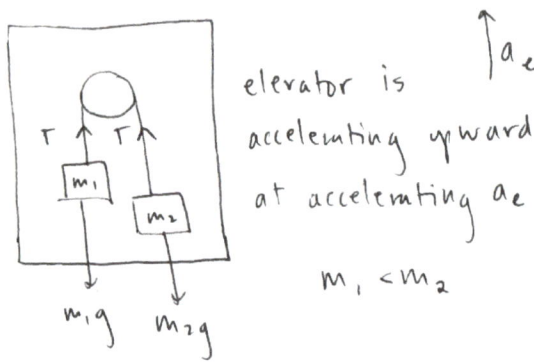

elevator is accelerating upwards at accelerating a_e

$m_1 < m_2$

KEY: Tension throughout pulley is equal
Remember elevator physics

- m_1: natural tendency to accelerate <u>upwards</u> bc $m_1 < m_2$
  * thus elevator a_e contributes to its acceleration

- m_2: natural tendency to accelerate downwards bc $m_2 > m_1$
  * thus elevator a_e acts against its acceleration

Remember: 2 masses are connected by same pulley thus they have same acceleration relative to the pulley a_p

Setup: Same as classic pulley problem but now take into account effect of elevator acceleration in net movement

m_1: $\quad T - m_1 g = m_1 (a_p + a_e)$

m_2: $\quad m_2 g - T = m_2 (a_p - a_e)$

_____

$\quad\quad m_2 g - m_1 g = m_1 (a_p + a_e) + m_2 (a_p - a_e)$

Lifting a Pulley of Masses

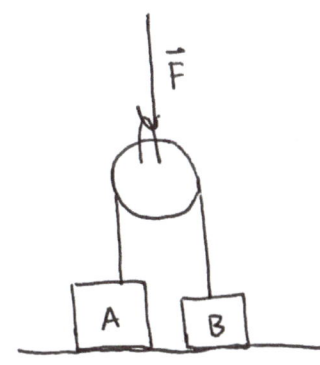

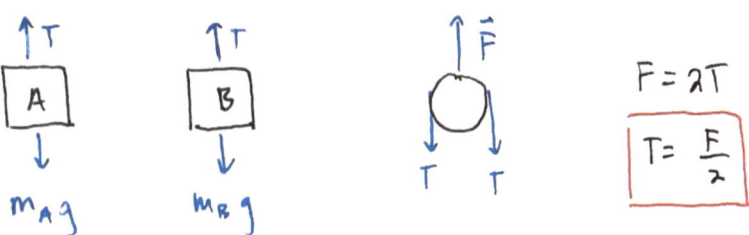

$m_A > m_B$ Blocks are initially at rest on the floor
Upward force $\vec{F}$ is applied to pulley

$F = 2T$

$\boxed{T = \dfrac{F}{2}}$

Finding acceleration of blocks requires 3 cases.

Case 1: $F < 2m_B g$

because $m_A > m_B$ and $F < 2m_B g$ → Tension is not enough to lift masses

$\boxed{a_A = a_B = 0}$

Case 2: $2m_B g < F < 2m_A g$ Key: Tension enough to lift B but not A

$T - m_B g = m_A a_B$ $\boxed{a_A = 0}$

$\boxed{a_B = \dfrac{F}{2m_B} - g}$

Case 3: $2m_A g < F$ Key: Tension enough to lift both masses

$T - m_A g = m_A a_A$ $T - m_B g = m_B a_B$

$\dfrac{F}{2} - m_A g = m_A a_A$ $\dfrac{F}{2} - m_B g = m_B a_B$

$\boxed{a_A = \dfrac{F}{2m_A} - g}$ $\boxed{a_B = \dfrac{F}{2m_B} - g}$

Inclines

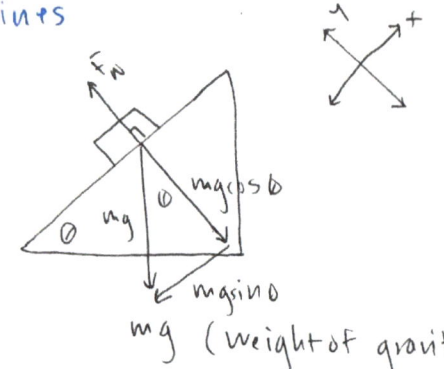

mg (weight of gravity)

$mg\cos\theta$ & $mg\sin\theta$ → tendency of object to move due to <u>gravity</u>

Key: tilt x & y axis so parallel to incline

No motion:

$$\boxed{F_N = mg\cos\theta}$$

$ma = mg\sin\theta$

$\boxed{a = g\sin\theta}$

<u>interesting</u>
→ acceleration does NOT depend on mass

With Friction

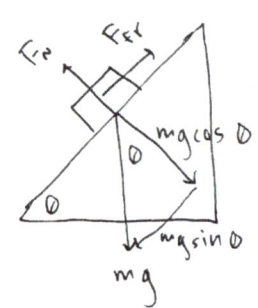

Key: Remember F_{Fr} always <u>opposes motion</u>
acts in opposite direction

x: $mg\sin\theta = F_{Fr}$ y: $F_N = mg\cos\theta$

$mg\sin\theta = \mu F_N$

$\cancel{m}g\sin\theta = \cancel{m}g\cos\theta \, \mu$

$\boxed{\mu = \tan\theta}$

*constant speed on ramp
$a = 0$ → $F = ma = 0$

Flat Turn WITH friction

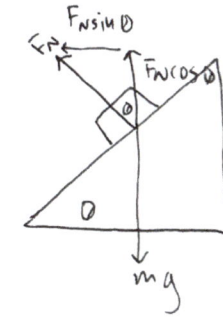

F_{fr} — points radially inward

$F_N = mg$ $F_{fr} = \dfrac{mv^2}{r}$

$\mu mg = \dfrac{mv^2}{r}$

$\mu g = \dfrac{v^2}{r}$

Banked Road (normal)

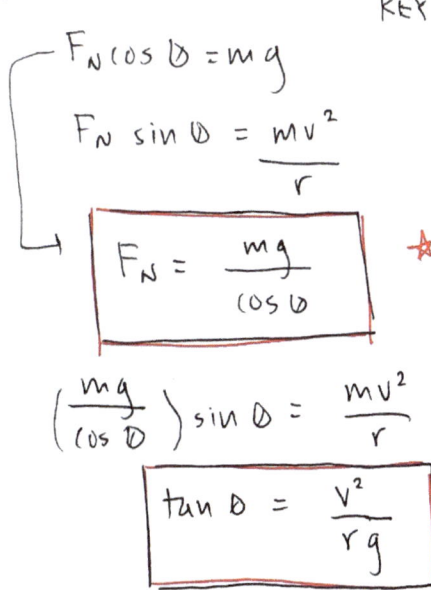

$F_N \cos\theta = mg$

$F_N \sin\theta = \dfrac{mv^2}{r}$

$$\boxed{F_N = \dfrac{mg}{\cos\theta}}$$

$\left(\dfrac{mg}{\cos\theta}\right)\sin\theta = \dfrac{mv^2}{r}$

$$\boxed{\tan\theta = \dfrac{v^2}{rg}}$$

KEY: horizontal component of N force is responsible for centripetal acceleration

★VERY IMPORTANT
- different Normal force as compared to incline

Incline Normal $= mg\cos\theta$

Banked Turns WITH FRICTION

nonskidding car → friction is static between tires and road
skidding car (dangerous) → friction is kinetic between tires and road
KEY: 2 scenarios   Friction serves as supporting force

① Car is going fast and thus has tendency to move up ramp
   Friction acts downwards

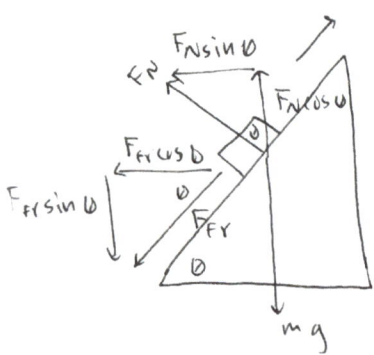

$y:\ F_N \cos\theta = F_{fr} \sin\theta + mg$

$F_N \cos\theta = \mu F_N \sin\theta + mg$

$F_N (\cos\theta - \mu\sin\theta) = mg$

$$\boxed{F_N = \frac{mg}{\cos\theta - \mu\sin\theta}}$$

Key: Solve for F_N

$x:\ F_{fr}\cos\theta + F_N \sin\theta = \frac{mv^2}{r}$

$$\boxed{\mu F_N \cos\theta + F_N \sin\theta = \frac{mv^2}{r}}$$

substitute F_N

② Car going too slow and thus has tendency to fall down ramp
   Friction acts upwards

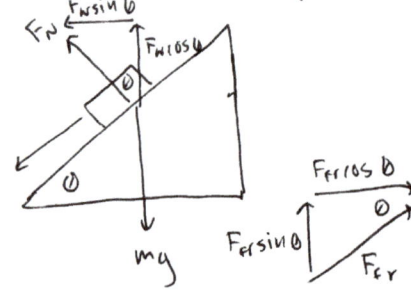

$y:\ mg = F_N \cos\theta + F_{fr}\sin\theta$

$mg = F_N \cos\theta + \mu F_N \sin\theta$

$mg = F_N (\cos\theta + \mu\sin\theta)$

$$\boxed{F_N = \frac{mg}{\cos\theta + \mu\sin\theta}}$$

be careful of θ placement on F_{fr}

$x:\ \frac{mv^2}{r} = F_N \sin\theta - \mu F_N \cos\theta$

Ramp + Pulley + 2 Boxes + Friction

Key: 2 scenarios

① Tendency for block m_1 to move down ramp and m_2 upwards

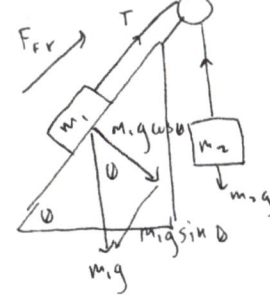

$m_1 a = m_1 g \sin\theta - F_{fr} - T$

$m_2 a = T - m_2 g$

___

$m_1 a + m_2 a = m_1 g \sin\theta - \mu m_1 g \cos\theta - m_2 g$

$$a = \frac{m_1 g \sin\theta - \mu m_1 g \cos\theta - m_2 g}{(m_1 + m_2)}$$

★ friction acts upward

② Tendency for block m_1 to move up ramp and m_2 downward

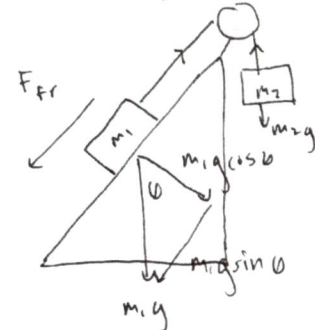

$m_1 a = T - \mu m_1 g \cos\theta - m_1 g \sin\theta$

$m_2 a = m_2 g - T$

___

$m_1 a + m_2 a = m_2 g - \mu m_1 g \cos\theta - m_1 g \sin\theta$

$$a = \frac{m_2 g - \mu m_1 g \cos\theta - m_1 g \sin\theta}{m_1 + m_2}$$

★ friction acts downward

2 Masses Stretched over a pulley incline NO friction

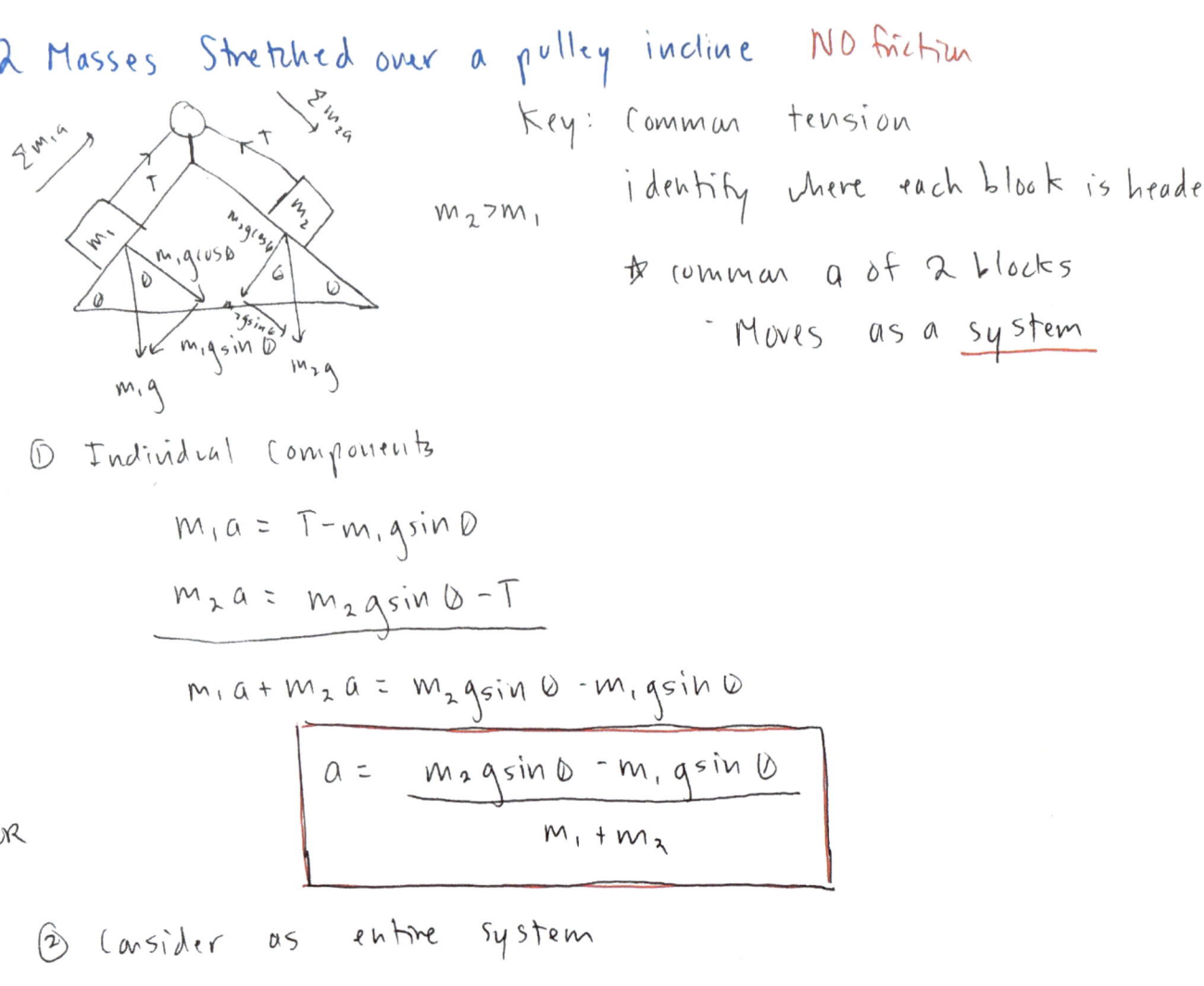

Key: Common tension

identify where each block is headed

$m_2 > m_1$

★ common a of 2 blocks
- Moves as a <u>system</u>

① Individual Components

$$m_1 a = T - m_1 g \sin\theta$$

$$m_2 a = m_2 g \sin\theta - T$$

$$m_1 a + m_2 a = m_2 g \sin\theta - m_1 g \sin\theta$$

$$\boxed{a = \frac{m_2 g \sin\theta - m_1 g \sin\theta}{m_1 + m_2}}$$

OR

② Consider as entire system

$$(m_1 + m_2)a_{tot} = m_2 g \sin\theta - m_1 g \sin\theta$$

$$\boxed{a_{tot} = \frac{m_2 g \sin\theta - m_1 g \sin\theta}{m_1 + m_2}}$$

2 connected masses rotating on a frictionless table   Period = P

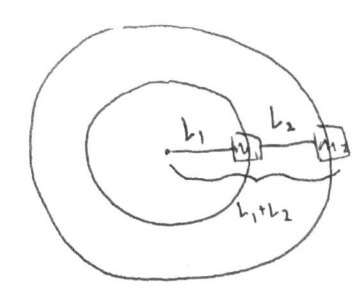

KEY: centripetal acceleration acting inward
* different tension in L_1 & L_2

m_1: $\quad \dfrac{m_1 v^2}{L_1} = T_1 - T_2$

m_2: $\quad T_2 = \dfrac{m_2 v^2}{L_1 + L_2} = \dfrac{m_2 \left(\dfrac{2\pi r}{P}\right)^2}{L_1 + L_2}$

Period = P $\rightarrow$ $v = \dfrac{2\pi r}{P}$

$\dfrac{m_1 v^2}{L_1} = T_1 - \dfrac{4\pi^2 m_2 (L_1 + L_2)}{P^2}$

$\boxed{T_2 = \dfrac{4\pi^2 m_2 (L_1 + L_2)}{P^2}}$

$\boxed{T_1 = \dfrac{m_1 v^2}{L_1} + \dfrac{4\pi^2 m_2 (L_1 + L_2)}{P^2}}$

Toss Rock up hill of a certain angle

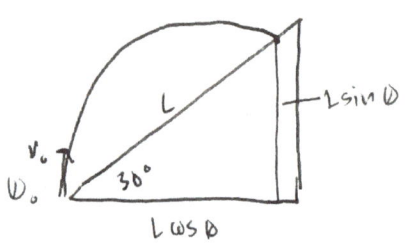

KEY: recognize x component of distance = $L\cos\theta$
y component = $L\sin\theta$

* throwing projectile angle is different from angle of hill

① $x = x_0 + v_0 t + \frac{1}{2}at^2$ ② $y = y_0 + v_0 t + \frac{1}{2}at^2$

$L\cos\theta = v_0 \cos\theta \, t$ $L\sin\theta = v_0 \sin\theta \, t - \frac{g}{2}t^2$

$$\frac{L\sin\theta}{L\cos\theta} = \frac{v_0 \sin\theta \, t - \frac{g}{2}t^2}{v_0 \cos\theta \, t}$$

$\downarrow 30$

$$\boxed{\tan\theta = \tan\theta_0 - \frac{gt}{2v_0 \cos\theta_0}}$$

Maximization / Minimization Strategy

1. Identify what variable you're maximizing / minimizing
2. Identify free variable that can be altered
3. Write maximization variable in terms of free variable
4. Take derivative w/ respect to free variable and set to 0 → solve for free variable

$\sin\theta = \sin(180 - \theta)$

* an object's change in velocity depends on not only the acceleration but also the time over which acceleration gets to act on object

* **Normal force counteracts vertical forces**

Tug of War

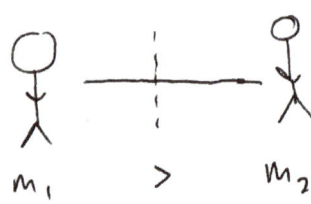

*Both experience same Tension in rope

Both experience the same force BUT because m_2 has less mass, it has a greater acceleration and thus reaches the midpoint faster

Dice swinging in accelerating car

car accelerates forward → dice swings back at certain angle

KEY: Σ (net force) NEVER appears in free body diagram (ma)

dice acceleration = car's acceleration → it is a RESULT (not an actual force)

★ Whenever one object is motionless with respect to another object
→ they "share" the same velocity and acceleration

$T\cos\theta = mg$

$T\sin\theta = ma$ → a of car

$$\frac{T\sin\theta}{T\cos\theta} = \frac{ma}{mg}$$

$$\boxed{\tan\theta = \frac{a}{g}}$$

3 Blocks Pushed Together

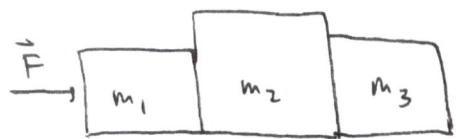

net force acting on entire system is F

whole system accelerates together

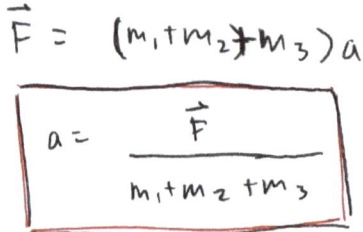

$$\vec{F} = (m_1 + m_2 + m_3)a$$

$$a = \frac{\vec{F}}{m_1 + m_2 + m_3}$$

① F → [F_{21}] ←   ② F_{12} → [ ]   ③ F_{32} ← [ ] F_{23} →

$F_{12} = -F_{21}$ $F_{32} = -F_{23}$ equal in magnitude but opposite direction

action reaction forces

Key: they all share same acceleration

① $F - F_{21} = m_1 a$

$F_{21} = F_1 - m_1 a$

③ $F_{23} = m_3 a$

Complex Pulley

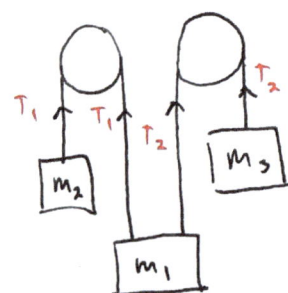

key: m_1 will accelerate down
m_2 and m_3 will accelerate up

$T_1 ↑ [m_2] ↓ m_2 g$ $T_1 ↑ ↑ T_2 [m_1] ↓ m_1 g$ $↑ T_2 [m_3] ↓ m_3 g$

$m_2 a = T_1 - m_2 g$

$m_1 a = m_1 g - T_1 - T_2$

$m_3 a = T_2 - m_3 g$

$$a = \frac{m_1 - m_2 - m_3}{m_1 + m_2 + m_3} g$$

Friction between 2 blocks
- Find max F so top block does NOT slide
key: cannot overcome static friction

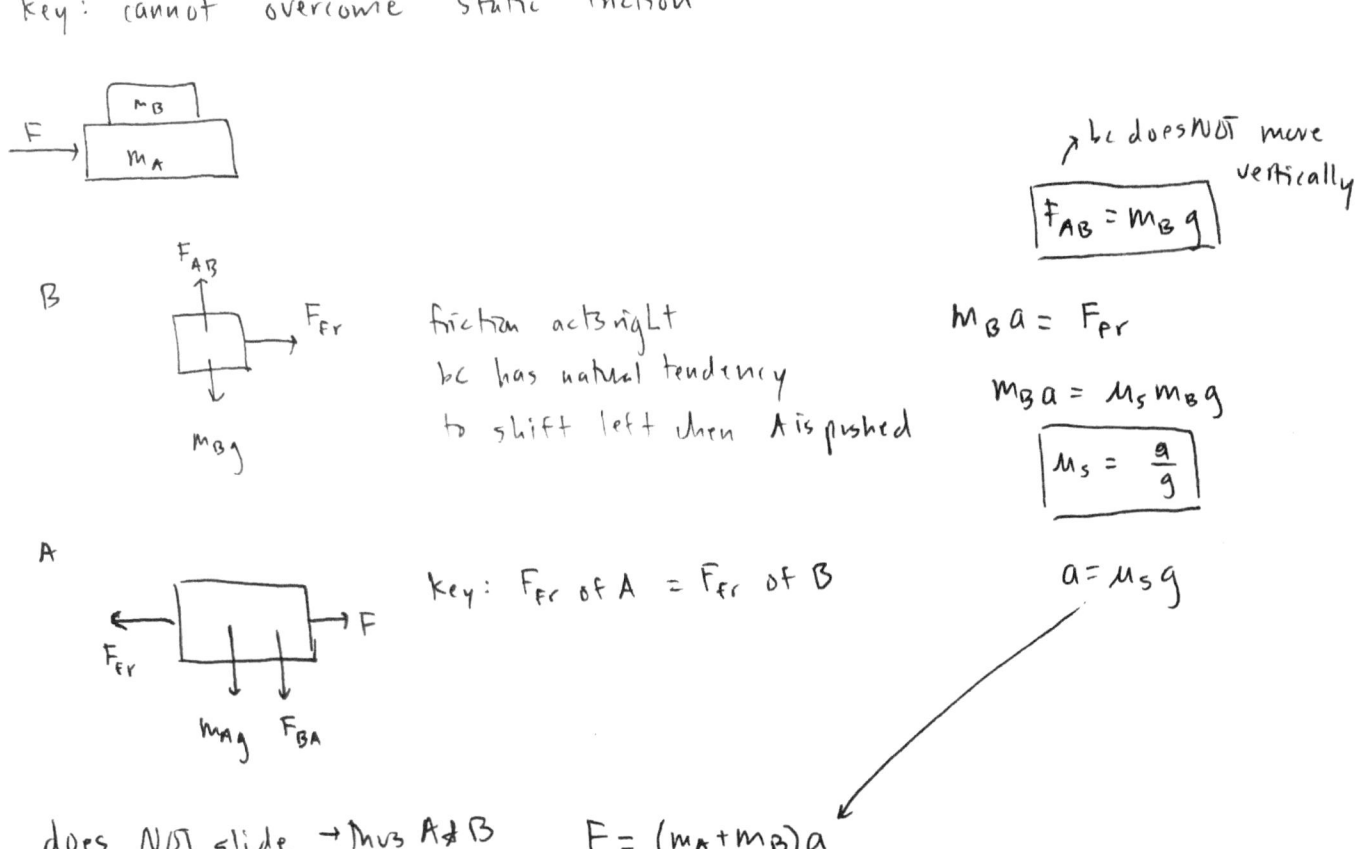

B: friction acts right bc has natural tendency to shift left when A is pushed

key: F_{fr} of A = F_{fr} of B

bc does NOT move vertically
$$F_{AB} = m_B g$$

$m_B a = F_{fr}$

$m_B a = \mu_s m_B g$

$$\mu_s = \frac{a}{g}$$

$a = \mu_s g$

does NOT slide → thus A & B move as a system

$F = (m_A + m_B) a$

$$F = (m_A + m_B) \mu_s g$$

2 Blocks on top on incline w/ pulley

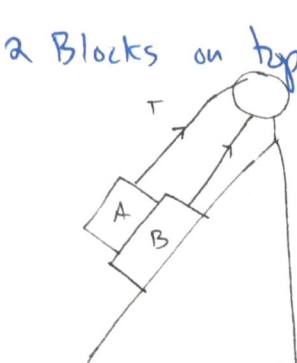

$mass_B > mass_A$

friction between B and ramp

★ Same rope has same tension

KEY: blocks do not stay stuck together (move as a system) thus do NOT consider $m_A + m_B$
   DON'T be REDUNDANT

★ Understand tension is consistent throughout the rope

Block A:

net movement is up ramp

② $\boxed{\begin{aligned} x&: m_A a = T - m_A g \sin\theta \\ y&: N_{BA} = m_A g \cos\theta \end{aligned}}$

Block B:

$N_{AB} = N_{BA}$

N_r — normal force exerted by ramp

net movement is down ramp

→ be careful w/ normal force

$\boxed{\begin{aligned} x&: m_B a = m_B g \sin\theta - T - F_{fr} \\ y&: 0 = N_r - N_{BA} - m_B g \sin\theta \end{aligned}}$

↓

★ $N_r = m_A g \cos\theta + m_B g \sin\theta$

① $m_B a = m_B g \sin\theta - T - \mu \underbrace{(m_A g \cos\theta + m_B g \sin\theta)}_{\mu N_r}$

trick: find equation to add to cancel out T

② $\underline{m_A a = T - m_A g \sin\theta}$

$m_B a + m_A a = m_B g \sin\theta - m_A g \sin\theta - \mu (m_A g \cos\theta + m_B g \sin\theta)$

$\boxed{a = \dfrac{m_B g \sin\theta - m_A g \sin\theta - \mu (m_A g \cos\theta + m_B g \sin\theta)}{(m_B + m_A)}}$

Remember: Normal force between block A and block B exert on each other are the same $F_{AB} = F_{BA}$

Normal Force between Ramp & B only account for mass B

→ DON'T BE REDUNDANT

Block Inside Inverted Rotating Cone

Given: Time for 1 rev = P

maintain constant height → find max & min of P

Very similar to car on banked turn with friction (be careful of θ)

Use geometry

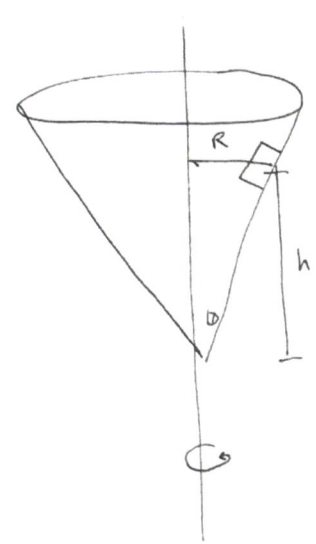

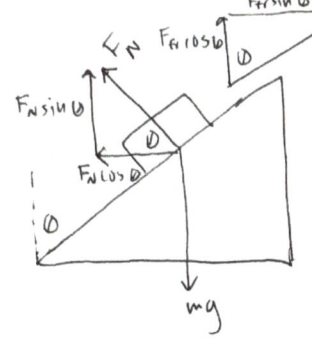

$\tan\theta = \dfrac{R}{h}$

$\boxed{R = h\tan\theta}$ radius of cone

① scenario 1 — block has tendency to slide **down** → friction acts upward

$V = \dfrac{2\pi r}{T} = \dfrac{2\pi r}{P}$

x: $\dfrac{mv^2}{r} = F_N\cos\theta - F_{fr}\sin\theta$

$\dfrac{m}{h\tan\theta}\left(\dfrac{2\pi r}{P}\right)^2 = F_N\cos\theta - \mu F_N\sin\theta$

y: $mg = F_N\sin\theta + F_{fr}\cos\theta$

$mg = F_N\sin\theta + \mu F_N\cos\theta$

$\boxed{F_N = \dfrac{mg}{\sin\theta + \mu\cos\theta}}$

$\dfrac{4\pi^2 m}{P^2}(\tan\theta) = F_N(\cos\theta - \mu\sin\theta)$

$\dfrac{4\pi^2 m}{P^2}\tan\theta = \left(\dfrac{mg}{\sin\theta + \mu\cos\theta}\right)(\cos\theta - \mu\sin\theta)$

$$\boxed{P^2 = \dfrac{4\pi^2 h}{g}\tan\theta\,\dfrac{\sin\theta + \mu_s\cos\theta}{\cos\theta - \mu_s\sin\theta}}$$

Scenario 2 — block moves up → friction acts down

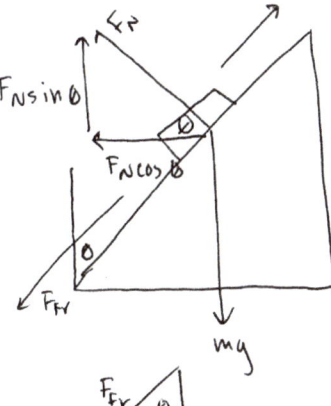

x: $\dfrac{mv^2}{r} = F_{fr}\sin\theta + F_N\cos\theta$

$\dfrac{mv^2}{r} = \mu F_N\sin\theta + F_N\cos\theta$

y: $mg + F_{fr}\cos\theta = F_N\sin\theta$

$mg + \mu F_N\cos\theta = F_N\sin\theta$ → solve for F_N

substitute into x equation

Elevator Physics

KEy: F_N is what scale reads

(diagram: F_N up, mg down)

① accelerating upwards

$$ma = F_N - mg$$
$$\boxed{F_N = ma + mg}$$

② accelerating downwards

$$ma = mg - F_N$$
$$\boxed{F_N = mg - ma}$$

*read 0 when cable breaks and is in free fall

③ decelerating upwards

$$-ma = F_N - mg$$
$$\boxed{F_N = mg - ma}$$

④ decelerating downwards

$$-ma = mg - F_N$$
$$\boxed{F_N = mg + ma}$$

Mass on either side

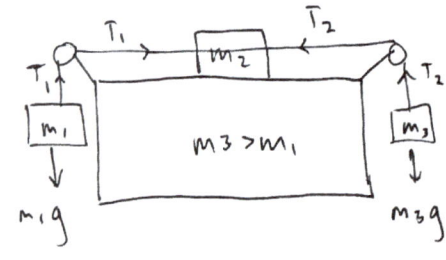

$m_3 > m_1$

* is tension throughout the string equal?

system: $(m_1 + m_2 + m_3) a = m_3 g - m_1 g$

$$\boxed{a = \frac{m_3 g - m_1 g}{m_1 + m_2 + m_3}}$$

★ $T_1 \neq T_2$

$$m_1 a = T_1 - m_1 g$$
$$m_3 a = m_3 g - T_2$$
$$m_2 a = T_2 - T_1$$
_____
$$(m_1 + m_2 + m_3) a = m_3 g - m_1 g$$

$$\boxed{a = \frac{m_3 g - m_1 g}{m_1 + m_2 + m_3}}$$

Tensions all cancel

2 Blocks Stacked + Pulley
Concept - Overcoming Static Friction

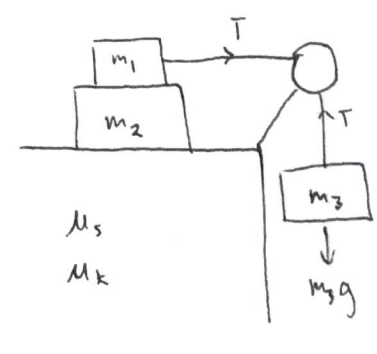

Key: Do not know whether m_1 & m_2 move together or move in opposite directions
- tension throughout string is the same

* if the tension created in pulley pulls with a greater force than static friction, then static friction turns into kinetic and the 2 blocks slide past each other

Scenario 1: Assume Move Together

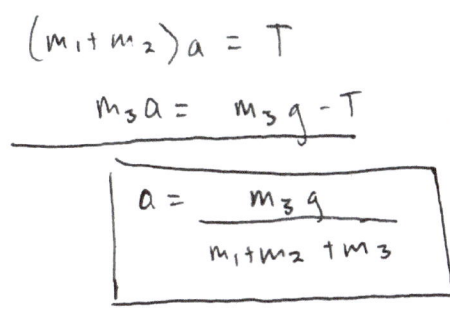

$(m_1 + m_2)a = T$

$m_3 a = m_3 g - T$

$$a = \frac{m_3 g}{m_1 + m_2 + m_3}$$

→ check to make sure does NOT create greater force than provided by static friction

all masses move w/ same acceleration

Scenario 2: M_1 & M_2 move separately → static friction turns into kinetic F_{fr}

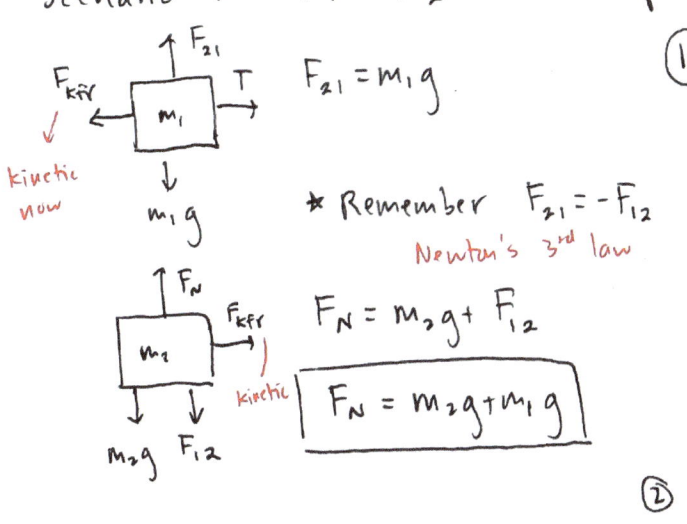

$F_{21} = m_1 g$

* Remember $F_{21} = -F_{12}$
Newton's 3rd law

$F_N = m_2 g + F_{12}$

$F_N = m_2 g + m_1 g$

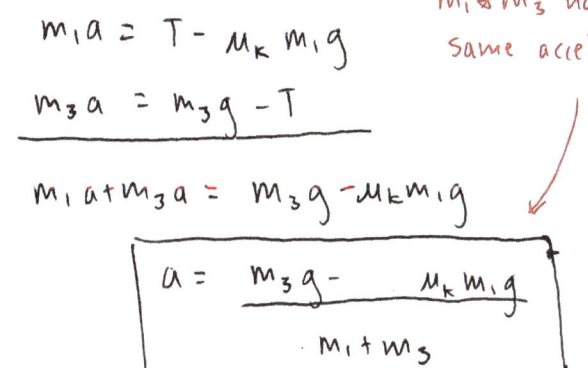

(1/3) $m_1 a = T - \mu_k m_1 g$

$m_3 a = m_3 g - T$

$m_1 a + m_3 a = m_3 g - \mu_k m_1 g$

$$a = \frac{m_3 g - \mu_k m_1 g}{m_1 + m_3}$$

m_1 & m_3 have same acceleration

(2) $m_2 a_2 = \mu_k (m_2 g + m_1 g)$

m_2 has different a

$$a_2 = \frac{\mu_k (m_2 g + m_1 g)}{m_2}$$

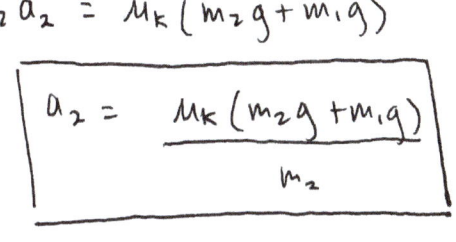

Moving Incline

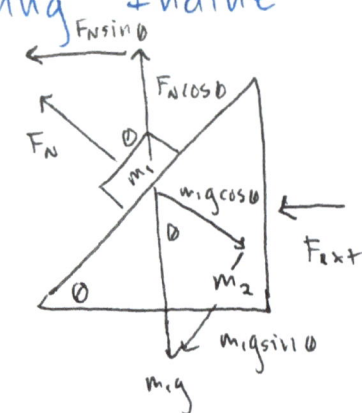

system: $F_{ext} = (m_1 + m_2) a$

Key: breaking up only normal force is necessary

take axis as $\hat{x}$ bc parallel to system movement

y: $F_N \cos\theta = m_1 g$

x: $F_N \sin\theta = m_1 a$ only net movement in x direction

$$\tan\theta = \frac{m_1 a}{m_1 g}$$

$a = \dfrac{F_{ext}}{m_1 + m_2}$ $\tan\theta = \dfrac{a}{g}$ $\boxed{F_{ext} = (m_1 + m_2) g \tan\theta}$

A Block in a Bowl

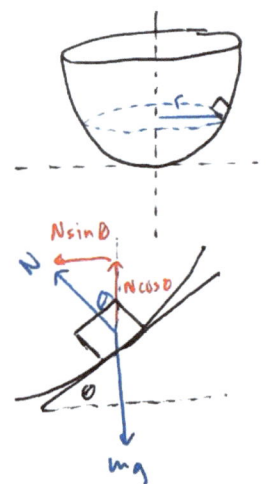

A block slides in a circle within a bowl. The speed of block is constant and no friction is present. How does the height, y, depend on radius, r, of the circle the block makes?

$\sum F_y$: $mg = N\cos\theta \rightarrow N = \dfrac{mg}{\cos\theta}$

$\sum F_x$: $\dfrac{mv^2}{r} = N\sin\theta \rightarrow \tan\theta = \dfrac{a}{g} = \dfrac{v^2}{rg}$

Key: slope $= \dfrac{dy}{dr} = \tan\theta = \dfrac{v^2}{rg} \rightarrow v = \sqrt{gr \dfrac{dy}{dr}}$

$v = \dfrac{2\pi r}{T}$ for one revolution

equate: $\sqrt{gr \dfrac{dy}{dr}} = \dfrac{2\pi r}{T}$

rearrange: $\dfrac{dy}{dr} = \dfrac{4\pi^2}{T^2} \dfrac{r}{g}$

integrate: $\boxed{y = \dfrac{2\pi^2}{gT^2} r^2}$

Cylinder cut out of a block — Find Max F so θ does NOT change

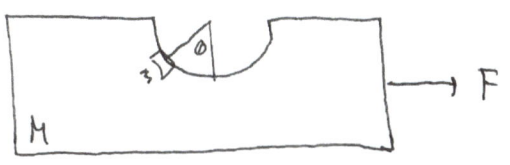

Remember inclined planes
keeping θ constant is equivalent
to keeping a car on a banked turn

Whole system moves
— thus shares same a

$F = (M+m)a$

2 scenarios — Block has tendency to slide down
— Thus Friction acts upward
— Block has tendency to slide up
— Friction acts downward

1st scenario

When 2 objects are motionless w/ respect to one another → share same V & a

$x: \quad F_N \sin θ - F_{fr} \cos θ = ma \quad$ accelerates w/ system

$y: \quad mg = F_{fr} \sin θ + F_N \cos θ \quad$ maintains constant height

---

$x: \quad F_N \sin θ - \mu F_N \cos θ = ma$

$\quad F_N (\sin θ - \mu \cos θ) = ma \quad \rightarrow \quad F_N = \dfrac{ma}{\sin θ - \mu \cos θ}$

$y: \quad mg = \mu F_N \sin θ + F_N \cos θ$

$\quad mg = F_N (\mu \sin θ + \cos θ) \quad \rightarrow \quad F_N = \dfrac{mg}{\mu \sin θ + \cos θ}$

$\dfrac{mg}{\mu \sin θ + \cos θ} = \dfrac{Fm}{(M+m)(\sin θ - \mu \cos θ)}$

$$\boxed{F_{min} = (M+m)g \, \dfrac{\sin θ - \mu_s \cos θ}{\cos θ + \mu_s \sin θ}} \qquad \boxed{F_{max} = (M+m)g \, \dfrac{\sin θ + \mu_s \cos θ}{\cos θ - \mu_s \sin θ}}$$

* difference between 2 scenarios is sign of μ_s

Car being towed up a ramp. Find greatest Distance it
Can be pulled up w/o breaking the cable

known: $M, \theta_1, \theta_2, T_{max}, t$

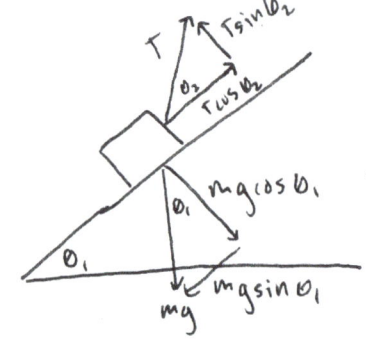

net movement is up ramp

x: $ma = T\cos\theta_2 - mg\sin\theta_1$

$$a = \frac{T\cos\theta_2}{M} - g\sin\theta_1$$

★use 1D kinematic equations

$x = x_0 + v_0 t + \frac{1}{2}at^2$

$$D = \frac{1}{2}\left(\frac{T\cos\theta_2}{M} - g\sin\theta_1\right)t^2$$

Vertically Connected Masses

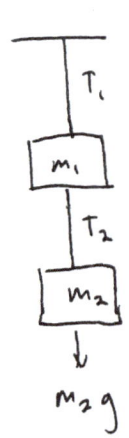

at Rest

$T_2 = m_2 g$

$T_1 = m_1 g + m_2 g$

accelerating upwards

$T_2 - m_2 g = am_2$

$$T_2 = m_2 a + m_2 g$$

$T_1 - (m_1 g + m_2 g) = (m_1 + m_2)a$

$$T_1 = a(m_1 + m_2) + g(m_1 + m_2)$$

Find Force Such that Small block does NOT slide up

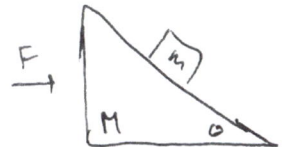

$F = (m+M)a$

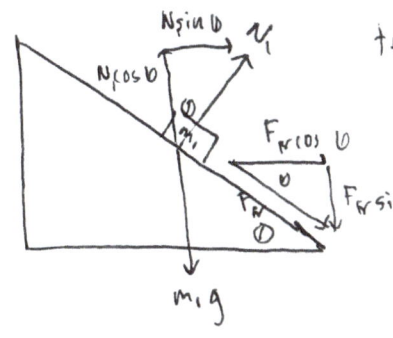

tendency to slide up
so friction acts down

$x: ma = N_1 \sin\theta + F_{fr}\cos\theta$

$y: mg + F_{fr}\sin\theta = N_1\cos\theta$

N_2 = balancing force between M and floor

$x: F - N_1\sin\theta - F_{fr}\cos\theta = Ma$

$y: Mg + N_1\cos\theta = F_{fr}\sin\theta + N_2$

friction acts opposite as m

* since blocks are in contact, every force has equal and opposite reaction

$my \rightarrow$ $$N = \frac{mg}{\cos\theta - \mu\sin\theta}$$

$mx \rightarrow$ $$a = \frac{\sin\theta + \mu\cos\theta}{\cos\theta - \mu\sin\theta} g$$

$$F = (m+M)\left(\frac{\sin\theta + \mu\cos\theta}{\cos\theta - \mu\sin\theta}\right)g$$

Several Blocks + Pulley

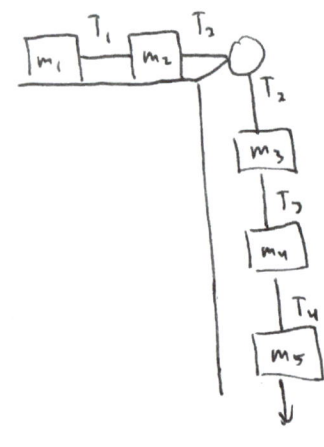

$m_1 a = T_1$

$m_2 a = T_2 - T_1$

$m_3 a = m_3 g + T_3 - T_2$

$m_4 a = m_4 g + T_4 - T_3$

$m_5 a = m_5 g - T_4$

$(m_1 + m_2 + m_3 + m_4 + m_5) a = m_3 g + m_4 g + m_5 g$

Find acceleration

$$a = \frac{m_3 g + m_4 g + m_5 g}{m_1 + m_2 + m_3 + m_4 + m_5}$$

Solve for T_3 — first solve for T_4

$T_4 = m_5 g - m_5 a \quad \rightarrow$ substitute

$m_4 a = m_4 g + m_5 g - m_5 a - T_3$

$$T_3 = m_4 g - m_4 a + m_5 g - m_5 a$$

substitute a

Crate sliding down 90° trough

find acceleration of crate

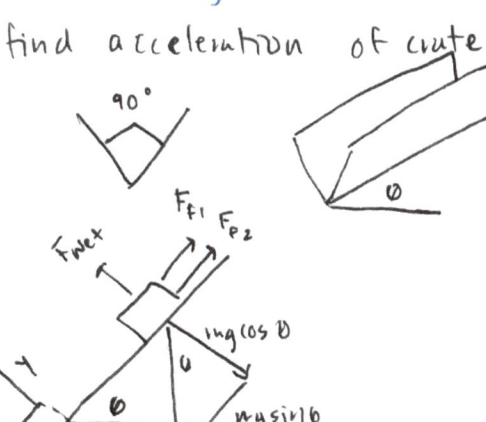
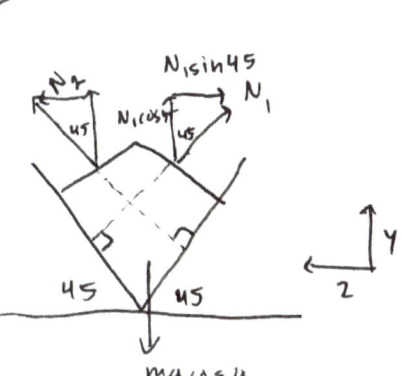

imagine z axis pointing out of block/ramp

$x:\ mg\sin\theta - F_{f1} - F_{f2} = ma_x$

$y:\ N_1\cos 45 + N_2\cos 45 - mg\cos\theta = ma_y = 0$

$z:\ ma_z = 0 = -N_1\sin 45 + N_2\sin 45$

$N_1 = N_2 = N$

$2N\cos 45 = mg\cos\theta$

$$N = \frac{mg\cos\theta}{2\cos 45}$$

$F_{f1} = F_{f2} = \mu_k N = F_f$

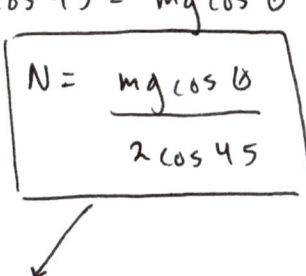

$ma_x = mg\sin\theta - 2F_f$

$ma_x = mg\sin\theta - \dfrac{\mu_k mg\cos\theta}{\cos 45°}$

$$a_x = g\sin\theta - \frac{\mu_k g\cos\theta}{\cos 45}$$

Ruby Riding With Truck

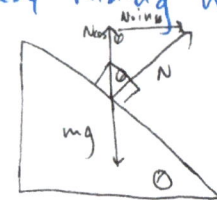

ruby is on the windshield - want it to not slide up or slide down
- has a wind resistance = $\frac{mg}{2}$

KEY: car is accelerating in x direction (carrying ruby with it)

ruby's total acceleration = $a_{rel\ to\ windshield}$ + $a_{of\ car}$

A = accelerating of car

$$|| \\ 0 \text{ bc does not slide up or down}$$

y: $N\cos\theta = mg$

x: $N\sin\theta - \frac{mg}{2} = mA$ → net x movement is car's acceleration

$\left(\frac{mg}{\cos\theta}\right)\sin\theta - \frac{mg}{2} = mA$

$\tan\theta\, mg - \frac{mg}{2} = mA$ → $\boxed{A = g\left(\tan\theta - \frac{1}{2}\right)}$

TRICK: if one object is inside another accelerating object, set the axises as parallel to movement of accelerating object for convenience

Block on an incline in a Moving Rocket Ship

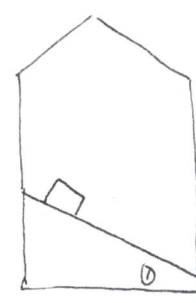

a. rocket sitting still → classic incline problem

$$ma = mg\sin\theta$$

$$\boxed{a = g\sin\theta}$$

b. rocket is ascending at constant speed

key: constant speed motion <u>adds nothing</u> to block's a

$$\boxed{a = g\sin\theta}$$

*remember elevator problem

c. rocket moves sideways at constant speed

key: constant speed motion <u>adds nothing</u> to block's a

$$\boxed{a = g\sin\theta}$$

d. rocket accelerating downwards

Key: ① you can choose either normal force or weight (mg) into x & y components depending on convenience

② if problem is inside accelerating system, set x & y coordinates parallel and ⊥ to direction of this acceleration

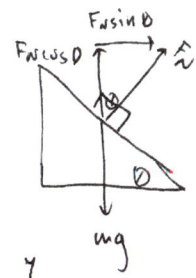

break up normal force instead of gravity

$$a_{block\ (total)} = a_{relative\ to\ floor} + a_{rocket}$$

*sum → just like walking in a moving train

X: $F_N \sin\theta = m(a_{rel}\cos\theta)$ no horizontal a of rocket

Y: $F_N \cos\theta - mg = -ma_{rel}\sin\theta - mA$ A = rocket

break up normal acceleration into x & y components

Uniform Circular Motion

- magnitude of velocity is constant
- direction of velocity is changing

radial acceleration - acts towards center of circle

velocity - acts in direction tangent to circle

$a \perp$ velocity

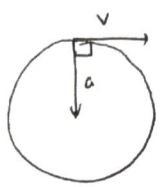

frequency = #rev/sec   period = time for 1 rev

$$\boxed{V = \frac{2\pi r}{T} = 2\pi r f}$$

→ set Σ(net) equal to that value

$\frac{mv^2}{r}$ is synonymous to ma application wise

Vertical Circle Rotation

vertical circle - NOT uniform circular motion bc gravity changes speed

BUT use equation anyway

★ KEY: remember centripetal force acts <u>radially inward</u>

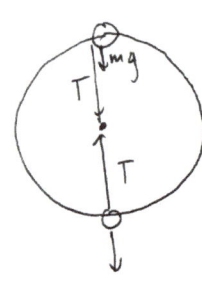

top

bottom

top: $\frac{mv^2}{r} = mg + T$ →

bottom: $T - mg = \frac{mv^2}{r}$

$$\boxed{\text{Slack} - T \text{ disappears} \\ \frac{mv^2}{r} = mg \\ V = \sqrt{rg}}$$

Conical Pendulum

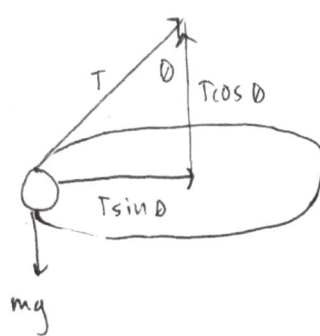

cord-length ℓ
radius $= \ell \sin\theta$

$T\cos\theta = mg$

$T\sin\theta = \dfrac{mv^2}{r}$

$T\sin\theta = \dfrac{mv^2}{\ell \sin\theta}$

Removable Floor — Thrill Ride

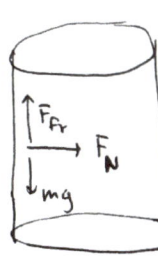

* F_{Fr} acts upwards against weight (mg) to keep person afloat

* F_N — acts as $\perp$ force wall is pushing on person

$F_{Fr} = mg$

$F_N = \dfrac{mv^2}{r}$

$\mu F_N = mg$

$\dfrac{\mu \cancel{m} v^2}{r} = \cancel{m} g \rightarrow \boxed{v = \sqrt{\dfrac{rg}{\mu}}}$

Rotating Object that Goes Slack At Top

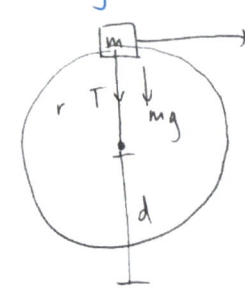

$$\frac{mv^2}{r} = \cancel{T} + mg \qquad \text{slack rope} \rightarrow \text{Tension} = 0$$

$$\frac{mv^2}{r} = mg$$

$$\boxed{v = \sqrt{rg}} \qquad \text{Remember: velocity in } \underline{x \text{ direction}}$$

☆ turns into projectile motion problem

y: $\quad -r - d = -\frac{g}{2} t^2$

$t = \sqrt{\dfrac{2r + 2d}{g}}$ time is the link between x & y components in projectile motion

x: $\quad d = vt$

$$\boxed{R = \sqrt{rg} \sqrt{\frac{2r + 2d}{g}}}$$

Crossing River

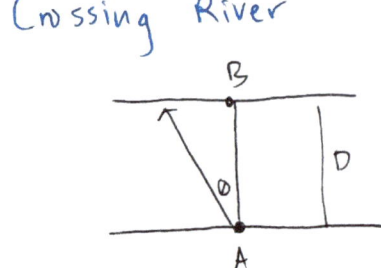

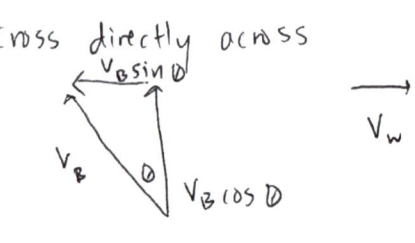

cross directly across

$V_B \sin\theta = V_w \qquad$ x-component is 0

$V_B \cos\theta \rightarrow$ y component directly across

$$\boxed{t = \frac{D}{V_B \cos\theta}}$$

☆ extra: to head across river <u>fastest</u> not just directly ahead, maximize y component of velocity

3 Blocks Don't get Confused

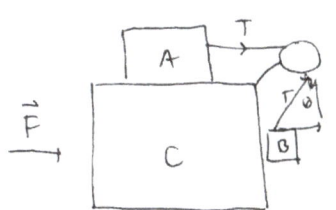

A does NOT move relative to C
find $\vec{F}$ (acts only on C)

A - acceleration for entire system

System: $\vec{F} = (m_A + m_B + m_C) A$

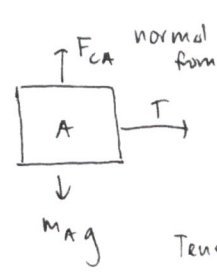

$F_{CA} = m_A g$

$T = m_A A$

Tension is only ΣF acting on A

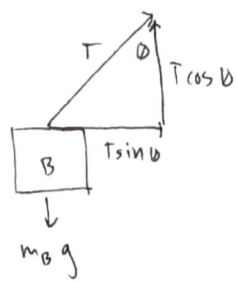

$T\cos\theta = m_B g$

$T\sin\theta = m_B A$

TRICK
identify there are trig functions involved → square both sides

substitute T

$(m_A A \cos\theta = m_B g)^2$

$(m_A A \sin\theta = m_B A)^2$

Sum

$m_A^2 A^2 \cos^2\theta = m_B^2 g^2$

$m_A^2 A^2 \sin^2\theta = m_B^2 A^2$

$\cos^2\theta + \sin^2\theta = 1$

$m_A^2 A^2 \underline{\cos^2\theta} + m_A^2 A^2 \underline{\sin^2\theta} = m_B^2 g^2 + m_B^2 A^2$

$A = \sqrt{\dfrac{m_B^2}{m_A^2 - m_B^2} g^2}$

$$\boxed{\vec{F} = (m_A + m_B + m_C)\sqrt{\dfrac{m_B^2}{m_A^2 - m_B^2} g^2}}$$

Weight of a Person in a Ferris Wheel

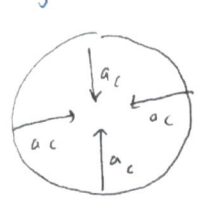
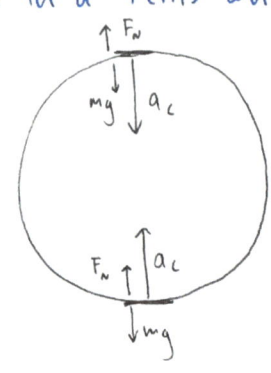

F_N – what scale reads

Top: $\dfrac{mv^2}{r} = mg - F_N$

$$\boxed{F_N = mg - \dfrac{mv^2}{R}}$$

Bottom: $\dfrac{mv^2}{r} = F_N - mg$

$$\boxed{F_N = mg + \dfrac{mv^2}{R}}$$

key: net acceleration is now centripetal $\left(\dfrac{mv^2}{R}\right)$

2 Blocks on a Ramp connected by a cord

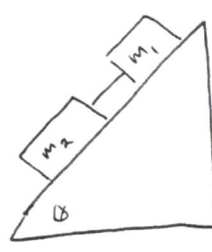

* system slides down ramp

key: F_{fr} opposed net direction of movement

① m_2

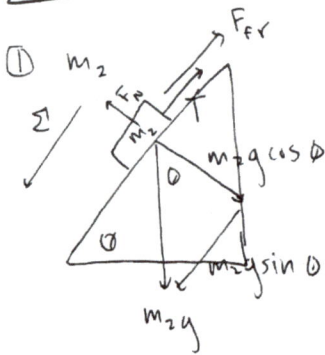

$m_2 a = m_2 g \sin\theta - T - \mu m_2 g \cos\theta$

②

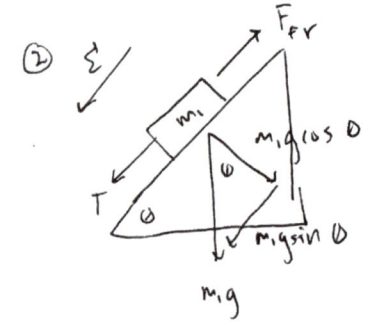

$m_1 a = m_1 g \sin\theta + T - \mu m_1 g \cos\theta$

Sum of equations: $m_2 a + m_1 a = m_2 g \sin\theta - \mu m_2 g \cos\theta + m_1 g \sin\theta - \mu m_1 g \cos\theta$

$$\boxed{a = \dfrac{m_2 g \sin\theta - \mu m_2 g \cos\theta + m_1 g \sin\theta - \mu m_1 g \cos\theta}{m_1 + m_2}}$$

* if considered as system, exact same thing except Tensions cancelled out

Carnival Ride

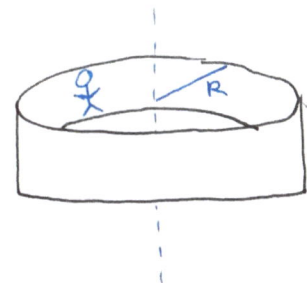
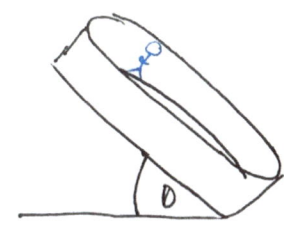

Rider stands against spinning cylindrical wall while it rotates. The floor drops so nothing supports feet of rider.

a. μ_s present between rider and wall
Find maximum period (T) to make one complete revolution

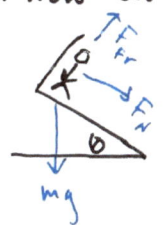

$$V = \frac{2\pi R}{T} \qquad F_N = \frac{mv^2}{R} \qquad F_{fr} = mg$$

$$F_{fr} = \mu_s F_N = \mu_s \frac{mv^2}{R} = \mu_s \frac{m 4\pi^2 R^2}{RT^2}$$

$$mg = \mu_s \frac{m 4\pi^2 R^2}{RT^2}$$

$$\boxed{T = \sqrt{\frac{\mu_s 4\pi^2 R}{g}}}$$

b. now consider tilted ride: find F_N when rider is at highest position

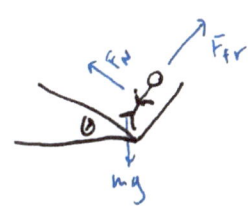

$$F_N + mg\sin\theta = \frac{mv^2}{R} \qquad F_{fr} = mg\cos\theta \text{ (static)}$$

$$\boxed{F_N = \frac{4\pi^2 mR}{T^2} - mg\sin\theta}$$

c. Find F_N at lowest position

$$F_N - mg\sin\theta = \frac{m 4\pi^2 R}{T^2}$$

$$\boxed{F_N = \frac{m 4\pi^2 R}{T^2} + mg\sin\theta}$$

d. Find force of static friction at height midway between highest and lowest positions

$$\boxed{F_{fr} = mg}$$

Varying Gravitation Along Earth's Surface

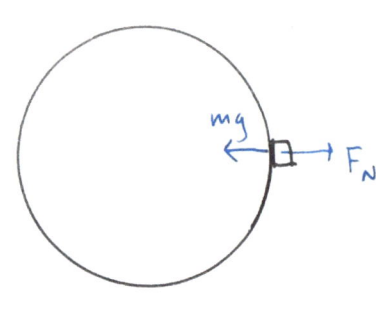

gravity at Earth's equator is different from gravity at Earth's other surfaces

Key: F_N = perceived gravitation

$$mg - F_N = \frac{mv^2}{R}$$

$$F_N = mg - \frac{mv^2}{R}$$

$$V = \frac{2\pi R}{T}$$

$$V^2 = \frac{4\pi^2 R^2}{T^2}$$

$$mg' = \frac{GM_E m}{R^2} - \frac{m \, 4\pi R}{T^2}$$

$$\boxed{g' = \frac{GM_E}{R^2} - \frac{4\pi R}{T^2}}$$

A person is released at a distance above Earth's surface. Find its velocity when reach Earth surface

Key: as person falls, force of gravity changes with distance from Earth's surface

Chain Rule

$$a = \frac{dv}{dt} = \frac{dv}{dr} \frac{dr}{dt}$$

$$a = v \frac{dv}{dr}$$

$$ma = -\frac{GM_E m}{r^2}$$

$$\frac{v \, dv}{dr} = -\frac{GM_E}{r^2}$$

$$v \, dv = -\frac{GM_E}{r^2} dr$$

$$\int_0^{v_f} v \, dv = \int_{2R_E}^{R_E} -\frac{GM_E}{r^2} dr$$

$$\tfrac{1}{2} v_f^2 = \frac{GM_E}{r_E} - \frac{GM_E}{2 r_E}$$

$$\boxed{v_f = -\sqrt{\frac{GM_E}{r_E}}}$$

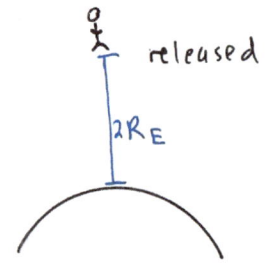

Escape Velocity

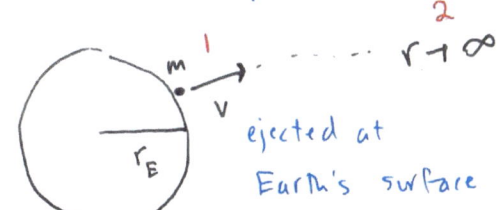

ejected at Earth's surface

Say you hate someone and want to get rid of them. Find the escape velocity at which you eject them so they never return back to earth.

$E_1 = U_1 + K_1$ $\qquad E_2 = U_2 + K_2$

$E_1 = \frac{1}{2}mv^2 - \frac{Gmm_E}{r_E}$ $\qquad E_2 = 0 + 0 = 0$ at $r \to \infty$

$E_1 = E_2$

$\frac{1}{2}mv^2 - \frac{Gmm_E}{r_E} = 0$

$$\boxed{v = \frac{2Gm_E}{r_E}}$$

Skier Leaving Face of Earth

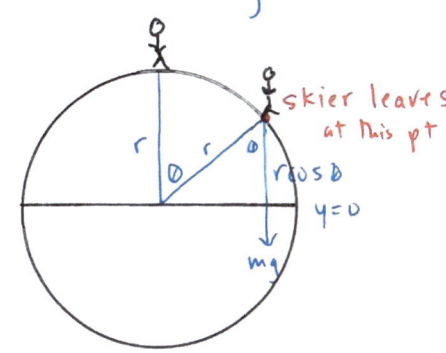

Key: Solve in terms of energy using middle of earth as reference point

$mgr = mgr\cos\theta + \frac{1}{2}mv^2$

all gravitational PE $\qquad$ gravitational PE + translational KE

$$\boxed{1. \quad v^2 = 2(gr - gr\cos\theta)}$$

at leaving point, skier loses contact with Earth's surface

thus $F_N = 0$

$-\cancel{F_N} + mg\cos\theta = \frac{mv^2}{r}$

$$\boxed{2. \quad v^2 = rg\cos\theta}$$

Note: can equate found equations 1 and 2

2 gliders collide with a Spring

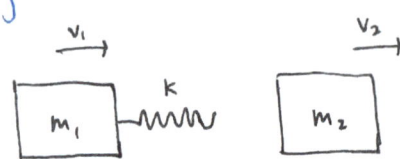

A mass moving at speed V_1 and a spring attached collides with another mass moving at speed V_2.

Key: The spring will compress until the velocities of each of the masses are the same.

Find speed of each mass at max. compression.

Find max compression: X_{max}

momentum conservation:

$$m_1 V_1 + m_2 V_2 = (m_1 + m_2) V_f$$

$$\boxed{V_f = \frac{m_1 V_1 + m_2 V_2}{m_1 + m_2}}$$

energy conservation:

$$\tfrac{1}{2} m_1 V_1^2 + \tfrac{1}{2} m_2 V_2^2 = \tfrac{1}{2} m_1 V_f^2 + \tfrac{1}{2} m_2 V_f^2 + \tfrac{1}{2} k X_{max}^2$$

substitution of V_f

$$\tfrac{1}{2} m_1 V_1^2 + \tfrac{1}{2} m_2 V_2^2 = \tfrac{1}{2} m_1 \left(\frac{m_1 V_1 + m_2 V_2}{m_1 + m_2} \right)^2 + \tfrac{1}{2} m_2 \left(\frac{m_1 V_1 + m_2 V_2}{m_1 + m_2} \right)^2 + \tfrac{1}{2} k X_{max}^2$$

$$k X_{max}^2 = \frac{m_1 m_2}{m_1 + m_2} (V_1 - V_2)^2$$

$$\boxed{X_{max} = \sqrt{\frac{1}{k} \frac{m_1 m_2}{m_1 + m_2} (V_1 - V_2)^2}}$$

Energy of Orbitting Satellites

KE:
$$\frac{mv^2}{r} = \frac{GMm}{r^2}$$

$$mv^2 = \frac{GMm}{r}$$

$$\tfrac{1}{2}mv^2 = \frac{GMm}{2r}$$

$$\boxed{KE = -\tfrac{1}{2}U}$$

Work to change orbit of satellite (from A → B)

$$W = (K_B - K_A) + (U_B - U_A)$$

$$\boxed{W_{net} = \tfrac{1}{2}(U_B - U_A)}$$

key: Force of Centripetal Acceleration = Force of Gravitational Attraction

Block + Incline + Spring

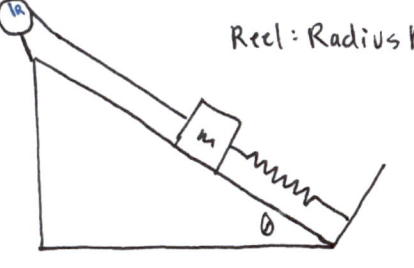

Reel: Radius R

A block of mass m is at rest on an incline at an angle θ to the horizontal. μ_k is present between block and incline. The block is connected to a reel at one end and a spring at the other. The spring is stretched at distance d from equilibrium. Find angular speed (ω) when block travels d back to equilibrium position.

key: Establish equilibrium position of spring to be zero for potential energy

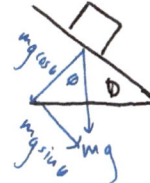

$E_i = \underset{\text{mass}}{mgd\sin\theta} + \underset{\text{spring PE}}{\tfrac{1}{2}kd^2}$

$E_f = \underset{\text{reel}}{\tfrac{1}{2}I\omega^2} + \underset{\text{mass}}{\tfrac{1}{2}mv^2} + \underset{\text{work done by friction}}{\mu_k mg\cos\theta \, d}$

energy conservation:

$mgd\sin\theta + \tfrac{1}{2}kd^2 = \tfrac{1}{2}I\omega^2 + \tfrac{1}{2}mv^2 + \mu_k mg\cos\theta \, d$ ← $v = \omega R$

$mgd\sin\theta + \tfrac{1}{2}kd^2 = \tfrac{1}{2}I\omega^2 + \tfrac{1}{2}m\omega^2 R^2 + \mu_k mg\cos\theta \, d$

$$\boxed{\omega = \sqrt{\frac{2mgd(\sin\theta - \mu_k\cos\theta) + kd^2}{I + mR^2}}}$$

Pushing a Block with Friction

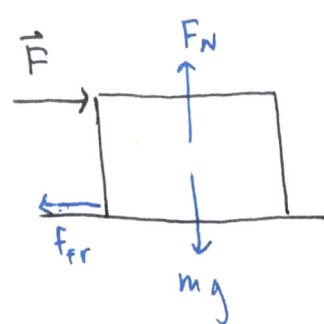

block of mass m is initially at rest
horizontal force, F, is applied for distance x

key: external force to block-table system = $\vec{F}$
internal force to block-table system = f_{fr}

external force creates external work

$$W_{ext.} = Fx$$

energy dissipated from friction

$$W_{fr} = \mu F_N = \mu mg$$

work of box = ΔKE = (Net force)x

$$\tfrac{1}{2}mv_{final}^2 - \tfrac{1}{2}mv_{initial}^2 = (F - f_r)x$$

$$\tfrac{1}{2}mv_{final}^2 = (F - \mu mg)x$$

$$v_{final} = \sqrt{\tfrac{2(F-\mu mg)x}{m}}$$

Relating Swinging Pendulum to Gravitational PE

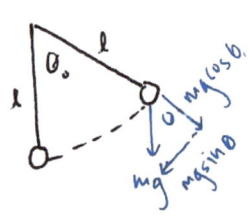

$$w = \int_0^{\theta_0} F \cdot ds$$

$$= \int_0^{\theta_0} F\ell \, d\theta$$

$$= \int_0^{\theta_0} mg\sin\theta \, \ell \, d\theta$$

$$= mg\ell \left[-\cos\theta\right]_0^{\theta_0}$$

$$w = mg\ell(1-\cos\theta_0)$$

$$\boxed{w = mgh}$$

$x = \ell\cos\theta$
$h = \ell - x$
$h = \ell - \ell\cos\theta$
$h = \ell(1-\cos\theta)$

Falling, Flexible Cord

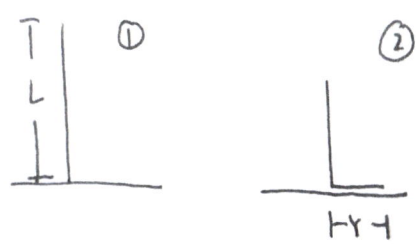

cord: Mass M, length L. Falls to the Table

Key: Use Energy

a. Find potential Energy $u(Y)$ of cord after Y length has fallen on the table

key: $u(Y)$ is a function of height, y

$$u = mgy$$
$$du = dm\, g\, y$$
$$\lambda = \frac{m}{y}$$
$$dm = \lambda\, dy$$

$\lambda = \frac{M}{L} \longrightarrow du = \lambda\, dy\, g\, y$

$$u(Y) = \int du = \frac{M}{L} g \int_0^{L-Y} y\, dy$$

$$u(Y) = \frac{Mg}{L} \frac{(L-Y)^2}{2}$$

$$\boxed{u(Y) = \frac{Mg}{2L}(L-Y)^2}$$

b. Find velocity $v(Y)$

Energy Conservation: $Y=0$: $u(Y=0)$ cord still at length L

at y: $E(Y) = u(Y) + \tfrac{1}{2} M v(Y)^2$

$$v(Y) = \sqrt{\frac{2(u(Y=0) - u(Y))}{M}}$$

$$= \sqrt{\frac{2}{M}\left(\frac{Mg}{2L} L^2 - \frac{Mg}{2L}(L-Y)^2\right)}$$

$$\boxed{v(Y) = \sqrt{\frac{g}{L}(L^2 - (L-Y)^2)}}$$

Collisions and Energy

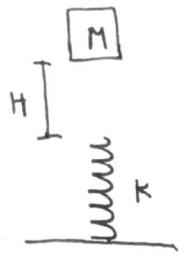

block of Mass M is dropped from height H onto spring. Block attaches to spring and compresses distance L before coming to rest.

work done by gravity: $W_g = MgL$

work done by spring: $W = -\frac{1}{2}kL^2$

speed of block just before it hits spring

Key: gravitational PE transformed into translational KE

$$MgH = \frac{1}{2}Mv^2$$

$$\boxed{v = \sqrt{2gH}}$$

Scenario 2:

$m = \frac{1}{4}M$

Just before block hits spring, bullet is fired into block. Find new max. compression of system. (L')

conservation of linear momentum

$$mv_B + M\sqrt{2gH} = (m+M)v_f$$

$$v_f = \frac{mv_B + M\sqrt{2gH}}{m+M} \quad \leftarrow m = \frac{1}{4}M$$

$$v_f = \frac{v_B + 4\sqrt{2gH}}{5}$$

→ neg: because below equilibrium position $y = 0$

conservation of energy

$$\frac{1}{2}(m+M)v_f^2 = \frac{1}{2}kL'^2 - (m+M)gL'$$

$$L' = \frac{\left(\frac{5M}{4}\right)g \pm \sqrt{\left(\frac{5M}{4}g\right)^2 + \frac{5M}{4}kv_f^2}}{k} \quad \text{quadratic formula}$$

$m + M = \frac{5M}{4}$

consider only pos. solution

$$\boxed{L' = \frac{\frac{5M}{4}g + \sqrt{\left(\frac{5M}{4}g\right)^2 + \frac{5M}{4}k\left(\frac{v_B + 4\sqrt{2gH}}{5}\right)^2}}{k}}$$

Testing Understanding of Collisions

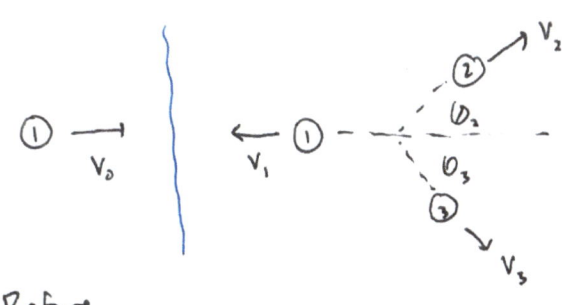

Before collision After Collision

Ball ① travels at $\vec{v}_0$ and collides with ② and ③ at rest.
All 3 balls have same mass but their diameters can be different.
Collisions are inelastic or elastic
frictionless

a. Can ball ① end up with velocity $\vec{v}_1 = -\vec{v}_0$?

  Key: final KE can never be greater than initial KE
  $$KE_i \geq KE_f$$

  if $v_1 = -v_0$ $\quad \frac{1}{2}mv_0^2 \geq \frac{1}{2}mv_1^2 + \frac{1}{2}mv_2^2 + \frac{1}{2}mv_3^2$

  $$0 \geq \frac{1}{2}mv_2^2 + \frac{1}{2}mv_3^2$$

  This states right side can either be $=0$ or negative

  NOT POSSIBLE   Balls must move after collision

b. If ball ① ends up at rest, could $|\theta_2| \neq |\theta_3|$?

  x: $mv_0 = mv_{2x} + mv_{3x}$ y: $0 = mv_{2y} - mv_{3y}$ ★ ball ① only has initial x component of velocity
  ★ $v_0 = v_{2x} + v_{3x}$ ★ $v_{2y} = v_{3y}$

  it is possible for $|\theta_2| \neq |\theta_3|$

c. If ball ① ends up moving backwards, can all 3 balls end up with the same speed but different directions?

  Thus, $|v_1| = |v_2| = |v_3| = v$

  x: $mv_0 = -mv_1 + mv_2\cos\theta_2 + mv_3\cos\theta_3$ y: $0 = mv_2\sin\theta_2 - mv_3\sin\theta_3$
  $v_0 = -v + v\cos\theta + v\cos\theta$ $0 = v\sin\theta_2 - v\sin\theta_3$
  $v_0 = (2\cos\theta - 1)v$ $\sin\theta_2 = \sin\theta_3$

  only requirement is $(2\cos\theta - 1) > 0$ $\theta_2 = \theta_3 = \theta$
  because v is positive

  thus, **possible**

Hidden Collision: Ball on a Slope

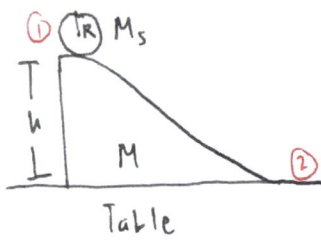

Table

Solid sphere of mass, M_s, and radius R is on top of a curved incline of mass M and height L.
There is no friction present.
The sphere rolls without slipping down and onto the table.
Find the final velocity of sphere and curved incline.

Key: When sphere rolls down, it pushes the curved incline backwards.

a. momentum conserved:
$$-MV_M + M_s V_s = 0$$
$$V_M = \frac{M_s}{M} V_s$$

note: incline has neg velocity bc moves in direction opposite to sphere

b. rolls w/o slipping:
$$\omega = \frac{V_s}{R}$$

c. Energy Conservation:
$$\underbrace{M_s g(h+R)}_{\text{gravitational PE to CM of sphere}} = \underbrace{M_s g R + \tfrac{1}{2} M_s V_s^2 + \tfrac{1}{2} I \omega^2}_{\text{translational + rotational KE of sphere}} + \underbrace{\tfrac{1}{2} M V_M^2}_{\text{translational KE of incline}}$$

plug a. & b. into c.
$$M_s g h = \tfrac{1}{2} M_s V_s^2 + \tfrac{1}{2} I \frac{V_s^2}{R^2} + \tfrac{1}{2} M \frac{M_s^2}{M^2} V_s^2$$

$$M_s g h = \tfrac{1}{2} M_s V_s^2 \left(1 + \frac{I}{M_s R^2} + \frac{M_s}{M}\right) \qquad I_{ball} = \tfrac{2}{5} M_s R^2$$

$$\boxed{V_s = \sqrt{\frac{2gh}{\tfrac{7}{5} + \frac{M_s}{M}}}}$$

$$\boxed{V_M = \frac{-M_s}{M} \sqrt{\frac{2gh}{\tfrac{7}{5} + \frac{M_s}{M}}}}$$

Hole in a Block

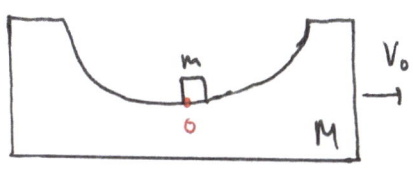

A block with a hole cut out of it slides across the table with velocity V_0.
A block of mass m is placed inside the large block. The small block will never fly out of large block.

a. Find max height, h, the small block will rise from bottom of hole.

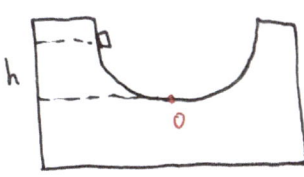

Key: When small block reaches max height, it is at a standstill and thus moves at some velocity as large block.

momentum conservation:

$P_i = P_f$

$MV_0 = (M+m)V_f$

$V_f = \dfrac{M}{M+m} V_0$

hidden collision

Energy Conservation

$\frac{1}{2} M V_0^2 = \frac{1}{2}(M+m)V_f^2 + mgh$

$h = \dfrac{V_0^2 M}{2mg}\left(1 - \dfrac{M}{M+m}\right)$

$$\boxed{h = \dfrac{V_0^2}{2g}\left(\dfrac{M}{M+m}\right)}$$

b. Find max horizontal velocity of small block

Key: The max velocity will be after it falls back from height h and passes again through the origin (pt 0)

momentum conservation: $MV_0 = mV_m + MV_M$

$V_M = \dfrac{MV_0 - mV_m}{M}$

energy conservation: $\frac{1}{2}MV_0^2 = \frac{1}{2}mV_m^2 + \frac{1}{2}MV_M^2$

$$\boxed{V_m = \dfrac{2V_0}{1 + \dfrac{m}{M}}}$$

Special Derivation: Derive Kinetic Energy Transfer Function for a Fully Elastic Collision

fully elastic collision — kinetic energy of a moving mass is <u>completely transferred</u> to a stationary mass during a head on collision

First: derive final velocities for any elastic head on collision

$$P: \quad m_A V_A + m_B V_B = m_A V_A' + m_B V_B'$$

$$E: \quad \tfrac{1}{2} m_A V_A^2 + \tfrac{1}{2} m_B V_B^2 = \tfrac{1}{2} m_A V_A'^2 + \tfrac{1}{2} m_B V_B'^2$$

rewrite:

$$P: \quad m_A (V_A - V_A') = m_B (V_B' - V_B)$$

$$E: \quad m_A (V_A^2 - V_A'^2) = m_B (V_B'^2 - V_B^2)$$

$$\downarrow$$

divide by momentum:

$$\frac{m_A (V_A - V_A')(V_A + V_A')}{m_A (V_A - V_A')} = \frac{m_B (V_B' - V_B)(V_B' + V_B)}{m_B (V_B' - V_B)}$$

$$\boxed{V_A - V_B = V_B' - V_A'} \quad \text{or} \quad V_A - V_B = -(V_A' - V_B')$$

$$V_B' = V_A - V_B + V_A' \quad \rightarrow \text{insert back into momentum equation}$$

$$m_A V_A + m_B V_B = m_A V_A' + m_B (V_A - V_B + V_A')$$

$$(m_A - m_B) V_A + 2 m_B V_B = (m_A + m_B) V_A'$$

$$\boxed{V_A' = V_A \left(\frac{m_A - m_B}{m_A + m_B} \right) + V_B \left(\frac{2 m_B}{m_A + m_B} \right)}$$

$$V_A' = V_B' - V_A + V_B \quad \rightarrow \text{insert back into momentum equation}$$

$$m_A V_A + m_B V_B = m_A (V_B' - V_A + V_B) + m_B V_B'$$

$$2 m_A V_A + (m_B - m_A) V_B = (m_A + m_B) V_B'$$

$$\boxed{V_B' = V_A \left(\frac{2 m_A}{m_A + m_B} \right) + V_B \left(\frac{m_B - m_A}{m_A + m_B} \right)}$$

continued →

Second: Find Kinetic Energy Transfer

① → ②
M_1 M_2 $M_2 = X M_1$

u = speed of masses before collision
v = speed after collision

f mass ① $KE_i = \frac{1}{2} m_1 u_1^2$ $KE_f = \frac{1}{2} m_1 v_1^2$

$$\Delta KE = \frac{1}{2} m_1 u_1^2 - \frac{1}{2} m_1 v_1^2$$

$$= \frac{1}{2} m_1 (u_1^2 - v_1^2)$$

change in energy of mass 1 is how much energy transferred to mass 2

Fractional decrease in KE

$$\frac{\Delta KE}{KE} = \frac{\frac{1}{2} m_1 (u_1^2 - v_1^2)}{\frac{1}{2} m_1 u_1^2} = 1 - \left(\frac{v_1}{u_1}\right)^2$$

using equation previously derived

$$v_1 = u_1 \left(\frac{m_1 - m_2}{m_1 + m_2}\right) + u_2 \left(\frac{2 m_2}{m_1 + m_2}\right)$$

initial speed of mass 2 is at rest $u_2 = 0$

$$v_1 = u_1 \left(\frac{m_1 - m_2}{m_1 + m_2}\right)$$

→ substitute into fractional KE eq.

$$\frac{\Delta KE}{KE} = 1 - \left(\frac{m_1 - m_2}{m_1 + m_2}\right)^2$$

$$\frac{\Delta KE}{KE} = \frac{4 m_1 m_2}{(m_1 + m_2)^2}$$

bc $m_2 = X m_1$

$$\boxed{f(x) = \frac{4x}{(1+x)^2}}$$

Note: less difference in masses, greater the transfer of KE
if $m_1 = m_2$ KE transfer = 100%

A ring rolls down a ramp and then inside a loop. It rolls without slipping. Find the height above the loop (h) such that the ring rolls all the way around the loop without falling off.

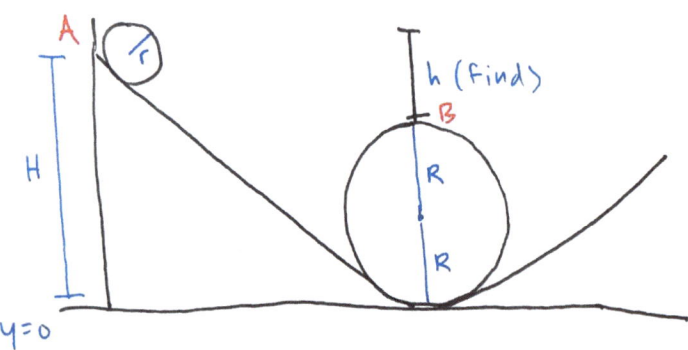

Conservation of energy

$$MgH = \tfrac{1}{2}Mv^2 + \tfrac{1}{2}I\omega^2 + Mg2R$$
$$\text{pt. A} \qquad\qquad \text{pt. B}$$

$$I = mr^2 \qquad \omega = \tfrac{v}{r}$$

$$\cancel{M}gH = \tfrac{1}{2}\cancel{M}v^2 + \tfrac{1}{2}\cancel{M}r^2\left(\tfrac{v^2}{r^2}\right) + \cancel{M}g2R$$

note: little r's cancel out

looking closer at pt. B

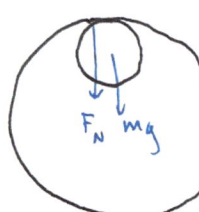

$$mg + \cancel{F_N} = \tfrac{mv^2}{r}$$

key: at extreme case where ring is about to fall off loop, the ring and loop will loose contact with each other. Thus, take $F_N = 0$

$$\boxed{v^2 = Rg}$$

$$gH = \tfrac{1}{2}v^2 + \tfrac{1}{2}v^2 + g2R$$
$$gH = v^2 + 2gR$$
$$gH = Rg + 2Rg$$
$$\boxed{H = 3R}$$

substitute

goal: to find h

$$h = H - 2R$$
$$h = 3R - 2R$$
$$\boxed{h = R}$$

Analyze all the forces acting on the sphere as it rolls down the track.

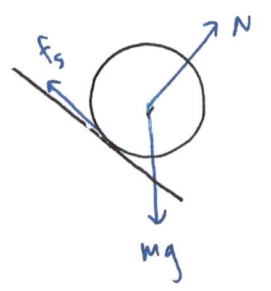

f_s: $w = 0$ — no distance over which f_s acts (pt of contact)

N: $w = 0$ — N force is ⊥ to displacement
$w = Nd\cos 90$

mg: $w = mgd\cos\theta$  $h = d\cos\theta$

2 Colliding Carts That Stick Together

initially at rest

a. Find velocities of all carts _immediately after collision_

   Key: 2 M's will stick and move together

   m will move along with same speed and slide up ▱

   important: all 3 masses do not yet move entirely as a system yet

$$Mv_0 = 2Mv_c$$

$$\boxed{v_c = \frac{v_0}{2}}$$

$$\boxed{\text{velocity of } m = v_0}$$

b. Find height mass m will reach on ▱

   Important: when mass m reaches max. height on ▱, it will pause momentarily and _move along_ with same velocity as system (all 3 blocks), _before it begins to slide back down_

   basically, m is at a standstill at height h

P: $mv_0 + Mv_0 = (m + 2M)v_s$

$$v_s = \frac{(m+M)v_0}{m + 2M}$$

E: $\frac{1}{2}mv_0^2 + \frac{1}{2}(2M)v_c^2 = mgh + \frac{1}{2}(m+2M)v_s^2$

$mv_0^2 + 2M\left(\frac{v_0}{2}\right)^2 = 2mgh + \frac{(m+M)^2}{(2M+m)}v_0^2$

$$\boxed{h = \frac{Mv_0^2}{4g(2M+m)}}$$

Fire Hydrants and Collecting Water

key: Analogous to Rocket Science - used derived formula

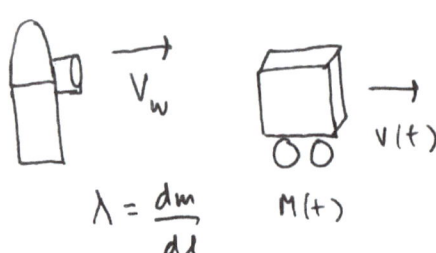

$\lambda = \dfrac{dm}{d\ell}$ $\quad M(t)$

A fire hydrant shoots water into a cart of initial Mass M_0. The water is shot with horizontal velocity V_w and λ.

key: Mass of cart increases as it collects water.

a. Find motion of the cart: $\dfrac{dv}{dt}$

already derived: $\quad \cancel{\dfrac{dP}{dt}} = M\dfrac{dv}{dt} - V_{rel}\dfrac{dM}{dt}$

$V_{rel} = V_w - V(t)$

*between water and cart

no external force

$$M(t)\dfrac{dv}{dt} = V_{rel}\dfrac{dM}{dt}$$

$$\boxed{\dfrac{dv}{dt} = \dfrac{1}{M(t)}(V_w - V(t))\dfrac{dM}{dt}}$$

b. Derive an expression for $\dfrac{dm}{dt}$ in terms of $\lambda, V(t), V_w$

Chain Rule: $\quad \dfrac{dm}{dt} = \dfrac{dm}{d\ell}\cdot\dfrac{d\ell}{dt} = \lambda\dfrac{d\ell}{dt}$

key: $\dfrac{d\ell}{dt} = V_{rel}$

* V_{rel} because referencing <u>moving</u> frame of the cart

$$\boxed{\dfrac{dm}{dt} = \lambda V_{rel} = \lambda(V_w - V(t))}$$

Classic Statics Problem

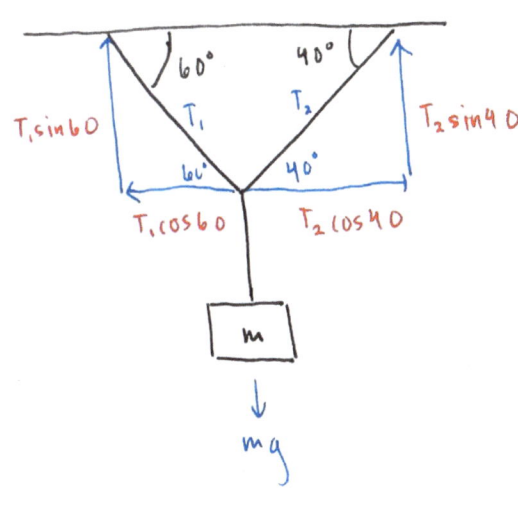

Key: $T_1 \neq T_2$

Divide T_1 & T_2 into x and y components.

For system to be in equilibrium, all forces must sum to 0.

$\sum F_x:$ $\quad T_1 \cos 60 = T_2 \cos 40$

$\sum F_y:$ $\quad mg = T_1 \sin 60 + T_2 \sin 40$

Classic Elastic Collision

Prove if a mass m with initial velocity V_o collides with another mass m then the angle between them after collision is $90°$.

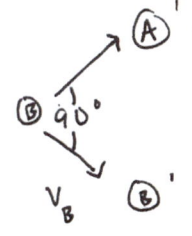

momentum conservation
$$m V_o = m V_A + m V_B$$
$$V_o = V_A + V_B$$

energy conservation
$$\tfrac{1}{2} m V_o^2 = \tfrac{1}{2} m V_A^2 + \tfrac{1}{2} m V_B^2$$
$$V_o^2 = V_A^2 + V_B^2$$

$V_o \cdot V_o = V_o^2$ dot product

$$V_o \cdot V_o = (V_A + V_B) \cdot (V_A + V_B)$$
$$V_o^2 = V_A^2 + V_B^2 + 2 V_A V_B$$
$$V_A^2 + V_B^2 = V_A^2 + V_B^2 + 2 V_A V_B$$
$$V_A \cdot V_B = 0$$

thus, $\boxed{\theta = 90°}$

key: masses must have same mass
must be elastic collision

2 ladders leaning against each other with a person on one ladder

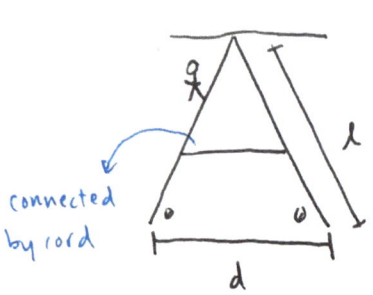
connected by cord

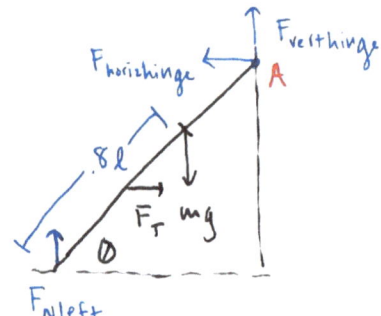
left ladder

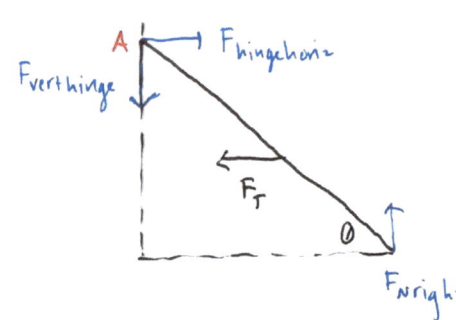
right ladder

$\cos\theta = \frac{\frac{1}{2}d}{\ell}$

$\theta = \cos^{-1}\left(\frac{d}{2\ell}\right)$

$\Sigma F_{vert}: F_{Nleft} + F_{verthinge} - F_{verthinge} + F_{Nright} - mg = 0$

$F_{Nleft} + F_{Nright} = mg$

left $\Sigma \tau: F_{Nleft}(\ell\cos\theta) - mg(.2\ell)\cos\theta - F_T(\frac{1}{2}\ell)\sin\theta = 0$

about pivot pt A

right $\Sigma \tau: -F_{Nright}(\ell\cos\theta) + F_T(\frac{1}{2}\ell)\sin\theta = 0$

Prevent Ladder From Sliding Down Wall

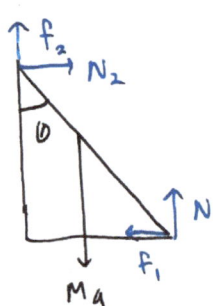

f - friction force to prevent sliding

$f_1 = \mu_s N_1$
$f_2 = \mu_s N_2$

$\Sigma F = 0$

x: $N_2 - f_1 = 0$
y: $N_1 + f_2 - Mg = 0$

→ algebra $N_1 = \frac{Mg}{1+\mu_s^2}$ $N_2 = \frac{\mu_s Mg}{1+\mu_s^2}$

$\Sigma \tau = 0$ $\frac{L}{2}\sin\theta\, Mg - L\cos\theta\, N_2 - L\sin\theta\, f_2 = 0$

→ substitution $\frac{\sin\theta}{2}Mg - \cos\theta\frac{\mu_s Mg}{1+\mu_s^2} - \sin\theta\frac{\mu_s^2 Mg}{1+\mu_s^2} = 0$

$\boxed{\tan\theta = \frac{2\mu_s}{1-\mu_s^2}}$ Max θ for ladder to not slide

Refrigerators, Trucks, and Pseudoforces

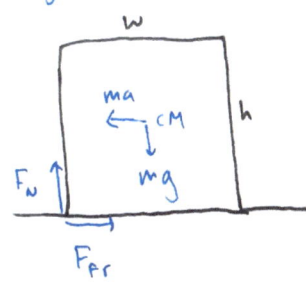

a refrigerator is on a truck that's accelerating to the right.

Key: ma = pseudo force, acts in direction <u>opposite to the truck</u>

ΣF_{horiz}: $F_{Fr} - ma = 0$

$F_{Fr} = ma$

ΣF_{vert}: $F_N - mg = 0$

$F_N = mg$

take torques <u>relative to center of mass</u>

F_N and F_{Fr} act on lower back corner to prevent fridge from tipping over

$\Sigma \tau$: $F_N(\tfrac{1}{2}w) - F_{Fr}(\tfrac{1}{2}h) = 0$

$$\frac{F_N}{F_{Fr}} = \frac{h}{w} = \frac{mg}{ma_{truck}} \qquad a = a_{truck}$$

$$\boxed{a_{truck} = g\,\frac{w}{h}}$$

Holding a Sign Up

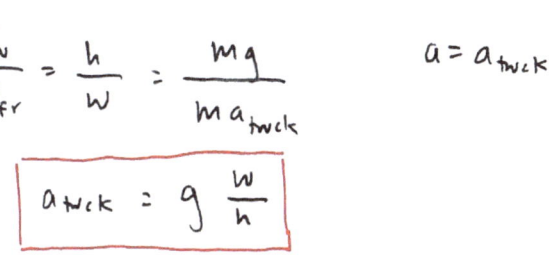

L = length of rod

$M = m_1 + m_2 \qquad \sin\theta = \dfrac{T_y}{T} \qquad \cos\theta = \dfrac{T_x}{T}$

$\Sigma F = 0 \qquad x: f_x - T_x = 0 \qquad y: f_y + T_y - Mg = 0$

$\qquad\qquad\qquad f_x = T_x$

$\Sigma \tau = 0 \qquad -\dfrac{L}{2}Mg + LT_y = 0$

$T_y = \dfrac{Mg}{2}$

$$\boxed{T = \dfrac{Mg}{2\sin\theta}}$$

James is moving a dresser...

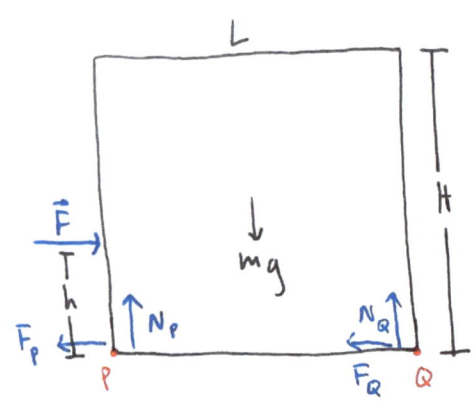

He applies force $\vec{F}$ at height h above ground.
μ_K is present between dresser and floor.
Dresser slides with constant velocity.

a. Express Force $\vec{F}$ in terms of m, g, μ_K

Key: constant velocity → zero acceleration

ΣF_x: $F = F_P + F_Q$ $\qquad$ ΣF_y: $N_P + N_Q = mg$

$F = \mu_K (N_P + N_Q)$

$$\boxed{F = \mu_K mg}$$

b. Find N_P and N_Q in terms of m, g, L, h, μ_K

ΣF: $N_P + N_Q = mg$ $\qquad$ $\Sigma \tau$: $mg \frac{L}{2} - Fh - N_P L = 0$ $\qquad$ key: torque taken about point Q

$$\boxed{N_Q = mg\left(\frac{1}{2} + \mu_K \frac{h}{L}\right)} \qquad \boxed{N_P = mg\left(\frac{1}{2} - \mu_K \frac{h}{L}\right)}$$

find one normal force using torque equation, then find second normal force using forces equation

c. Find max height (h_{max}) James can push dresser without it toppling over.

Key: Extreme case of toppling over:
Pt P begins to lose contact w/ ground
thus take $N_P = 0$

$\Sigma \tau = 0$ $\qquad$ key: torque taken about pt. Q

$mg \frac{L}{2} - Fh = 0$

$h = mg \frac{L}{2} \frac{1}{F}$ $\qquad$ using F found in part a.

$$\boxed{h_{max} = \frac{1}{2\mu_K}}$$

Finding Center of Mass of Cone

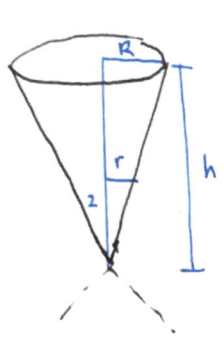

Key: by symmetry $x_{cm} = 0$ and $y_{cm} = 0$

find z_{cm}

$$z_{cm} = \frac{1}{M}\int z\, dm \qquad M = \int dm$$

$$= \frac{\int z\, dm}{\int dm} \qquad dm = \rho\, dV$$

$$dV = \pi r^2\, dz$$

$$= \frac{\int z \rho\, dV}{\int \rho\, dV} \qquad \frac{r}{z} = \frac{R}{h} \quad \text{apply similarity of } \Delta\text{'s}$$

$$r = \frac{zR}{h}$$

$$= \frac{\rho \int z \pi r^2\, dz}{\rho \int \pi r^2\, dz}$$

$$= \frac{\rho \pi \int z \left(\frac{zR}{h}\right)^2 dz}{\rho \pi \int \left(\frac{zR}{h}\right)^2 dz} = \frac{\rho \pi (R/h)^2 \int z^3\, dz}{\rho \pi (R/h)^2 \int z^2\, dz} = \frac{3}{4}h$$

$$\boxed{CM: (0, 0, \tfrac{3}{4}h)}$$

Center of Mass of Semicircle

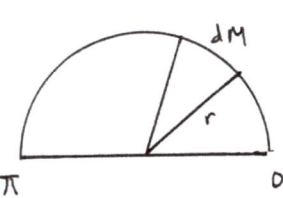

Key: by symmetry $x_{cm} = 0$

find y_{cm}

$$y_{cm} = \frac{1}{M}\int y\, dm \qquad dm = \frac{M}{\pi r}\, d\ell \quad d\ell = r\, d\theta$$

$$= \frac{1}{M}\int_0^\pi r \sin\theta \frac{M}{\pi}\, d\theta \qquad = \frac{M}{\pi r} r\, d\theta$$

$$= \frac{M}{\pi}\, d\theta$$

$$y_{cm} = \frac{2r}{\pi}$$

$$\boxed{CM: (0, \tfrac{2r}{\pi})}$$

Torques on a Spool of String

Case 1: Vertical Force

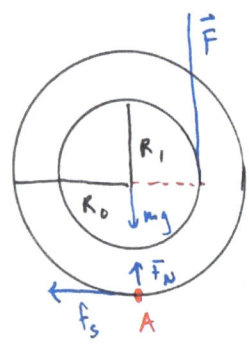

$\boxed{\tau = F R_1}$

spool will rotate in counter clockwise direction about pt. A

Key: Take torque to be relative to pt. of (pt A) contact with ground
→ simplifies calculations for τ
→ f_s won't be considered
→ mg won't be considered
→ F_N won't be considered

Case 2: Horizontal Force

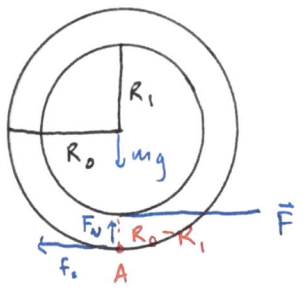

$\boxed{\tau = F(R_0 - R_1)}$

spool rotates clockwise

Case 3: Pulled at an angle

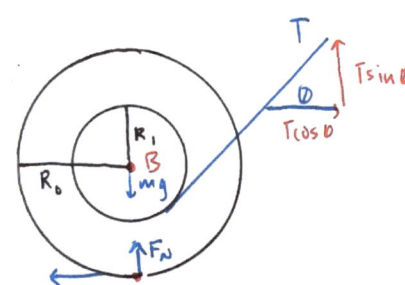

$\Sigma F_x:\ T\cos\theta = \mu_K N$

$\Sigma F_y:\ T\sin\theta + F_N = mg$

$\Sigma \tau:\ \mu_K N R_0 = T R_1$

Key: Torque is now taken about the center (B)

Rat Exercising on a Rotating Wheel

Treadmill: M_0, R_0, ω_0

Rat: $M_1 = \frac{1}{2} M_0$, runs at constant speed (stationary to wheel)

a. Find angular velocity (ω) after it grabs on

$L_i = L_f$

$M_0 R_0^2 \omega_0 = (M_0 R_0^2 + M_1 R_1^2)\omega_f$

$\boxed{\omega_f = \frac{4}{5}\omega_0}$

Angular Momentum Conserved

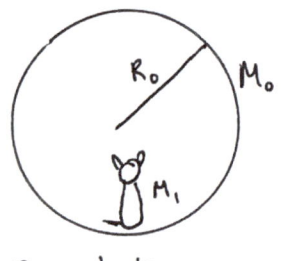

R_1 - rat to rotation axis

b. Energy when Rat Reaches Top of treadmill

Energy Conserved

$\frac{1}{2}(M_0 R_0^2 + M_1 R_1^2)\omega_f^2 = M_1 g 2 R_1$

Pushing Ball over Step

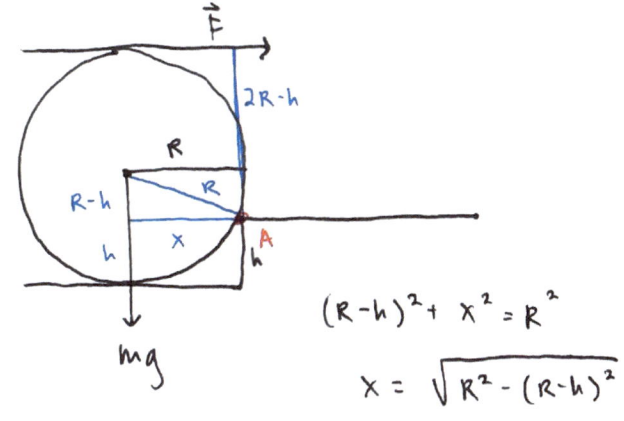

Key: all torques are taken about point of contact with step (pt A)

Find the lever distance as ⊥ component of each force relative to A

Forces acting on system
1. weight of ball
2. F (Force applied)

$$\sum \tau: \quad F(R-h) - mg\sqrt{R^2-(R-h)^2} = 0$$

If Force was applied to top of ball

to get over step

$$F(2R-h) - mg\sqrt{R^2-(R-h)^2} \geq 0$$

Note: opposite signs for each force because opposite torques created

$(R-h)^2 + x^2 = R^2$

$x = \sqrt{R^2-(R-h)^2}$

Bicycle Wheel Attached with Unknown Mass m

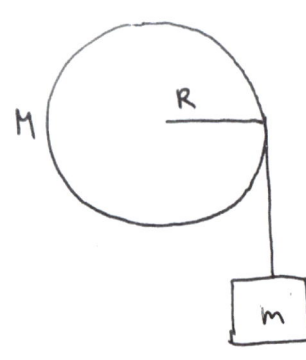

Releases mass m and measures time it takes for block to fall distance L. Find Mass m knowing L, t, R, m, and g.

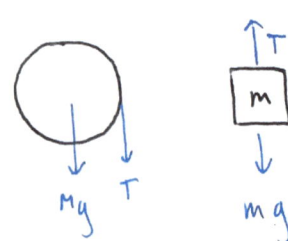

$\Sigma F: \quad ma = mg - T$

$\Sigma \tau: \quad I_{wheel}\, \alpha = TR \qquad I_{wheel} = MR^2$

key: Because rope is attached at <u>edge of wheel</u>, linear acceleration of edge of wheel = acceleration of the block.

$\alpha = aR$

$mg - Ma = ma$

$M = m\left(\dfrac{g}{a} - 1\right)$

using kinematics: $L = \tfrac{1}{2}at^2$

→ $a = \dfrac{2L}{t^2}$

substitute a

$\boxed{M = m\left(\dfrac{gt^2}{2L} - 1\right)}$

Unwinding a Spool of String

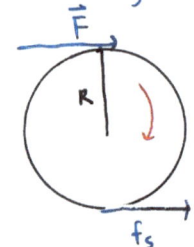

A spool of mass M and Radius R is unwound under constant force F.

$\Sigma F: \quad F + f_s = Ma$

$\Sigma \tau: \quad FR - f_s R = I\alpha \qquad a = R\alpha \quad I = \dfrac{MR^2}{2}$

key: f_s acts in same direction as Force applied because it creates a torque opposite to that of $\vec{F}$

static friction = resistance to movement
thus opposes rolling

$F + f_s = Ma$
$-F + f_s = -\tfrac{1}{2}Ma$
────────────
$2f_s = \tfrac{1}{2}Ma$

$f_s = \tfrac{1}{4}Ma$

$\boxed{f_s = \dfrac{F}{3} \text{ to the right}}$

$F + f_s = Ma$
$F - f_s = \tfrac{1}{2}Ma$
────────────
$2F = \tfrac{3}{2}Ma$

$\boxed{a = \dfrac{4}{3}\dfrac{F}{M} \text{ to the right}}$

$a = R\alpha$

$\boxed{\alpha = \dfrac{4}{3}\dfrac{F}{RM}}$

Pulleys and Torque on an Incline

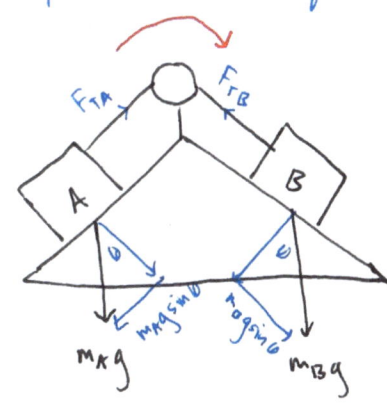

A: $F_{TA} - m_A g \sin\theta = m_A a$

B: $m_B g \sin\theta - F_{TB} = m_B a$

Key: tangential acceleration of pulley (a) is equal to acceleration of blocks
- tension on each side of string is <u>NOT equal</u>

$\Sigma \tau = (F_{TB} - F_{TA}) R$ net effect of tensions

$I\alpha = (F_{TB} - F_{TA}) R$ causes torque

Pulleys and Torque

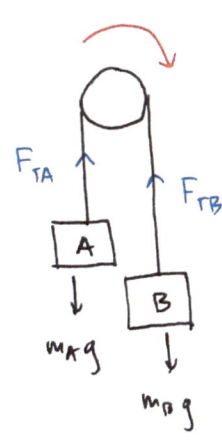

Key: $F_{TA} \neq F_{TB}$

A: $F_{TA} - m_A g = m_A a_{tang}$

B: $m_B g - F_{TB} = m_B a_{tang}$

Pulley: $\tau = (F_{TB} - F_{TA}) R = I\alpha$ $a = R\alpha$

Remember: acceleration of masses equals tangential acceleration of pulley

Angular Momentum of an Atwood Machine

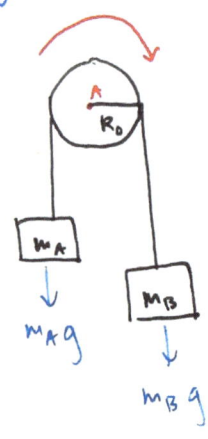

$$L = (m_A + m_B)vR_0 + I\frac{v}{R_0}$$

$\underbrace{}_{\text{masses}}$ $\underbrace{\phantom{I\frac{v}{R_0}}}_{\text{pulley}}$

$$\sum \tau = m_B g R_0 - m_A g R_0 \quad \text{torque about pt A}$$

$$\tau = \frac{dL}{dt}$$

$$(m_B - m_A)gR_0 = (m_A + m_B)R_0 \frac{dv}{dt} + \frac{I}{R_0}\frac{dv}{dt}$$

$\underbrace{}_{\text{derivative taken w/ respect to time}}$

$$\boxed{(m_B - m_A)gR_0 = (m_A + m_B)R_0 a + \frac{I}{R_0}a}$$ solve for a

Rotational Collision of Two Disks

A disk is dropped onto a rotating disk of angular frequency ω. Find angular frequency ω_f after collision

$$I_1 \omega = (I_1 + I_2)\omega_f$$

$$\boxed{\omega_f = \frac{I_1 \omega}{I_1 + I_2}}$$

Initial KE $= \frac{1}{2}I_1 \omega^2$ \qquad Final KE $= \frac{1}{2}(I_1 + I_2)\omega_f^2$

Drum + Pulley

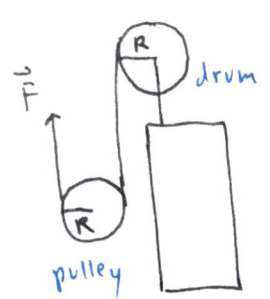

A string is wound around a drum and pulley of same Radius R. Find F pulled upwards such that pulley's CM doesn't move.

drum: $TR = I\alpha_d$

a. $TR = MR^2 \alpha_d$

pulley: $\Sigma \tau$: $TR - FR = I\alpha_p$

b. $TR - FR = \frac{1}{2}MR^2 \alpha_p$

ΣF: $T + F = Mg$ c.

Key: Relate α of drum and pulley

$\alpha_d = -\alpha_p \rightarrow$ b.

a. $\rightarrow$ $FR - TR = \frac{1}{2}MR^2 \alpha$

$TR = 2FR - 2TR$

$\boxed{T = \frac{2}{3}F}$ $\rightarrow$ c. $\rightarrow$ $\boxed{F = \frac{3}{5}Mg}$

Complex Fulcrum Problem

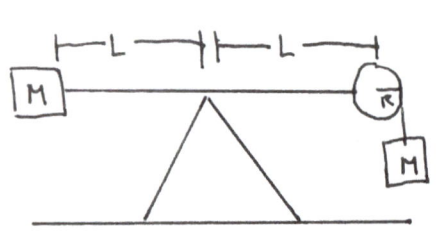

A rod of length $2L$ is pivoted at its center. A block of mass M is attached at one end. A pulley of rope is attached at other end with a block of same mass dangling.

a. Explain the dynamics of the system once released from equilibrium.

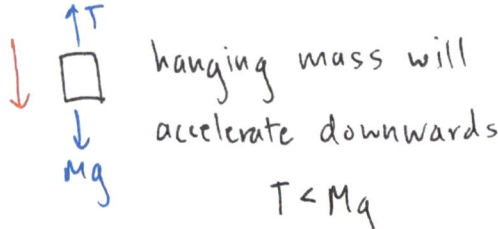

hanging mass will accelerate downwards
$T < Mg$

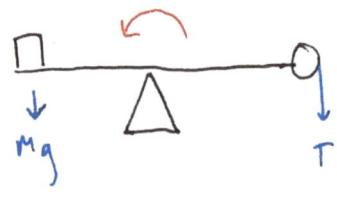

Rod will rotate counter clockwise
$T < Mg$

b. Find Tension in the rope when system is released

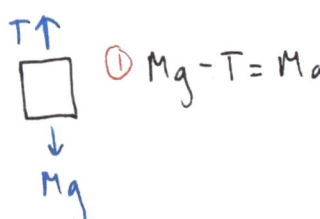

① $Mg - T = Ma$

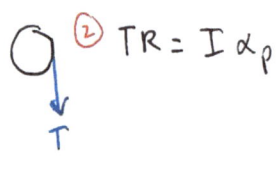

② $TR = I\alpha_p$

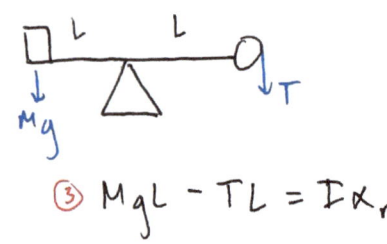

③ $MgL - TL = I\alpha_r$

hardest equation to derive : find relationship between a (angular acceleration) of rod and pulley

④ $a = \alpha_p R - \alpha_r L$

Hard algebra w/ 4 found equations:

$$T = \frac{Mg}{1 + \frac{MR^2}{2I}}$$

$$a = \frac{g}{1 + \frac{2I}{MR^2}}$$

Note: This problem shows how to consider the rotational dynamics of the pulley and rod separately. Then find the <u>link</u> between each of their accelerations.

Angular Momentum of a Rod Collision

① Inelastic collision

A putty flies and attaches itself to a vertical rod. The rod plus the putty falls over.

Key: The putty rod complex creates a new center of mass. The system then <u>rotates about this new center of mass</u>.

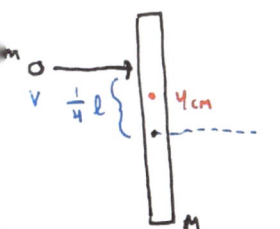

new center of mass

$$y_{cm} = \frac{m(\tfrac{1}{4}\ell) + M(0)}{m+M}$$

$$\boxed{y_{cm} = \frac{m\ell}{4(m+M)}}$$

above old CM

distance from clay to new CM:

$$y_{clay} = \tfrac{1}{4}\ell - y_{cm}$$

$$y_{clay} = \frac{M\ell}{4(m+M)}$$

conservation of linear momentum
velocity of old (cm)
— moves in a straight line

$$mv = (m+M)v_f$$

$$v_f = \frac{mv}{m+M}$$

conservation of angular momentum

$$L_{initial} = L_{final}$$

$$mv\, y_{clay} = (I_{rod} + I_{clay})\omega_f$$

$$\omega_f = \frac{mv\, y_{clay}}{\tfrac{1}{2}M\ell^2 + M\left[\frac{m\ell}{4(m+M)}\right]^2 + m\, y_{clay}^2}$$

parallel axis thm to new y_{cm}

$$\boxed{\omega_f = \frac{12mv}{\ell(7m+4M)}}$$

② **Elastic Collision** - putty continues to move along its line of path without sticking to the rod

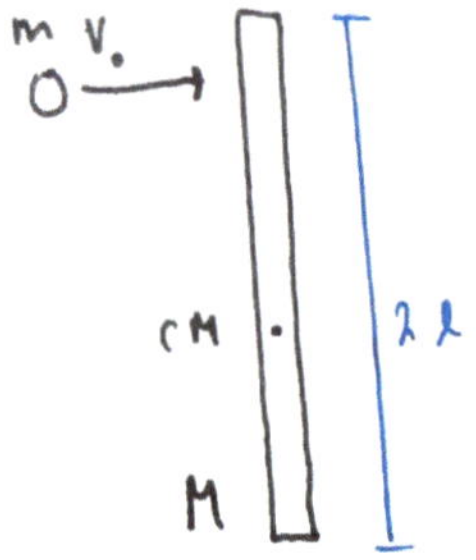

linear momentum conserved
$$mv_o = mv_f + Mv_{cm}$$
$$v_{cm} = v_o - v_f$$

angular momentum conserved
$$mv_o \ell = I_{cm} \omega_f + mv_f \ell$$

energy conserved
$$\tfrac{1}{2}mv_o^2 = \tfrac{1}{2}I_{cm}\omega_f^2 + \tfrac{1}{2}mv_f^2 + \tfrac{1}{2}Mv_{cm}^2$$

another scenario: now what if rod is fixed at end
rotation occurs around end of rod

Key: Linear Momentum is not conserved

Angular Momentum conserved
$$mv_o(2\ell) = I\omega_f + mv_f(2\ell)$$

Energy is conserved
$$\tfrac{1}{2}mv_o^2 = \tfrac{1}{2}mv_f^2 + \tfrac{1}{2}I\omega_f^2$$

note: use rod I for end of rod (not about its center)

2 Monkeys Stick to the ends of a Rod

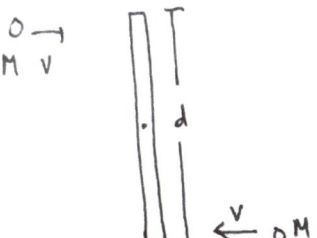

2 Monkeys, Mass M and speed v, simultaneously hit the ends of a rod of also Mass M and length d.

It is an <u>inelastic collision</u>.

a. Find the angular momentum <u>before the collision</u>

key: respect to center of rod

$$L = I\omega = mr^2 \frac{v}{r} = mrv$$

$r = \frac{d}{2} \rightarrow L = \frac{1}{2}mv$ for each particle

for both $\rightarrow$ $\boxed{L = dmv}$ like the car place!

b. Angular momentum <u>after the collision</u>

Key: Is there a net external torque acting on the system?

NO $\rightarrow$ L is conserved $\boxed{L = dmv}$

c. Find angular speed after collision

$$I = I_{rod} + 2I_{monkeys} = \left(\frac{1}{12} + \frac{6}{12}\right)Md^2 = \frac{7}{12}Md^2$$

$$2I_{monkeys} = 2Mr^2 = 2M\left(\frac{d}{2}\right)^2 = \frac{1}{2}Md^2$$

$$L = I\omega$$

$$\omega = \frac{L}{I} = \frac{dmv}{\frac{7}{12}md^2} \rightarrow \boxed{\omega = \frac{12}{7}\frac{v}{d}}$$

d. Find lost KE

initial KE = translational KE of each monkey = $2\left(\frac{1}{2}Mv^2\right) = Mv^2$

final KE = rotational KE of system = $\frac{1}{2}\left(\frac{7}{12}Md^2\right)\left(\frac{12}{7}\frac{v}{d}\right)^2 = \frac{6}{7}Mv^2$

$KE_{lost} = KE_f - KE_i = \boxed{\frac{1}{7}Mv^2}$

Pulley, 2 Masses, and Angular Momentum

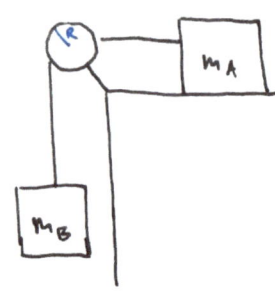

total angular momentum $(L) = L_A + L_B + L_{pulley}$

$= m_A R v + m_B R v + I \frac{v}{R}$

$\omega = \frac{v}{R}$

net torque is provided by gravity $= m_B g R$ (weight of block B)

key: take derivative of L with respect to time to find acceleration

$\Sigma \tau = \frac{dL}{dt}$

$m_B g R = \frac{d}{dt}\left[m_A R v + m_B R v + I \frac{v}{R} \right]$

$m_B g R = m_A R \frac{dv}{dt} + m_B R \frac{dv}{dt} + \frac{I}{R} \frac{dv}{dt}$

$m_B g R = m_A R \underline{a} + m_B R \underline{a} + \frac{I}{R} \underline{a}$

Then solve for a

Yoyo

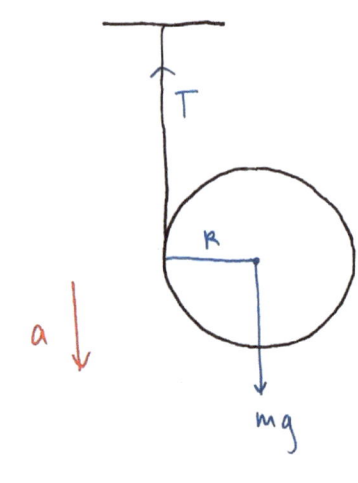

$\Sigma \tau = TR = I\alpha$

$TR = \frac{1}{2} M R^2 \alpha$

$T = \frac{1}{2} M R \alpha$

$\Sigma F: \quad mg - T = ma$

$mg - \frac{1}{2} M R \alpha = ma$

$\frac{mg}{R} - \frac{\frac{1}{2} M R \alpha}{R} = \frac{m R \alpha}{R}$

key: yoyo accelerates downward

$a = R\alpha$

$\frac{3}{2} \alpha = \frac{g}{R}$

$\boxed{a = \frac{2}{3} g}$

Tilted Angular Momentum

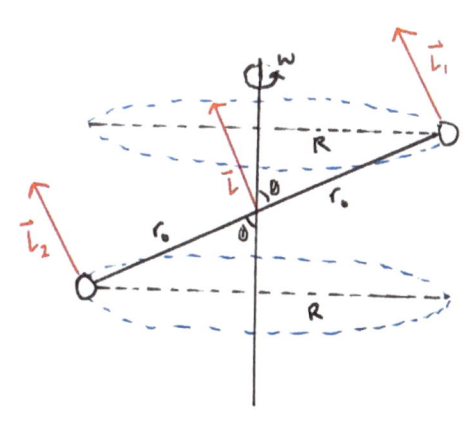

$\sin\theta = \dfrac{R}{r_0}$

$R = r_0 \sin\theta$

$L_1 = m_1 \vec{r}_1 \times \vec{v}_1$

$L_1 = m r_0 v = m r_0 R \omega \qquad v = R\omega$

$L_1 = m r_0^2 \sin\theta \, \omega$

$L_2 = m r_0^2 \sin\theta \, \omega \qquad$ symmetric

$L = L_1 + L_2$

$\boxed{L = 2 m r_0^2 \sin\theta \, \omega}$

The Spinning Top

A balanced top spins about its symmetric axis.

precession - motion created when a torque produces a <u>change in direction of rotation axis</u>

The axis will make on <u>angle ϕ</u> to the z axis and sweep out a cone as the top rotates.

To change change angular momentum, a torque is required.

The torque, in this scenario, is provided by the <u>weight of the top</u>.

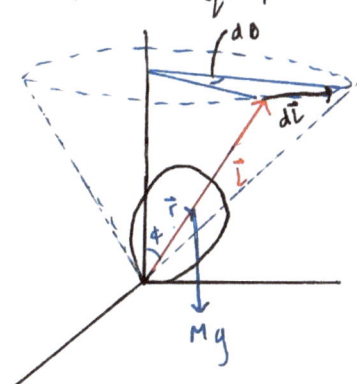

$dL = L \sin\phi \, d\theta$

$d\theta = \dfrac{dL}{L \sin\phi}$

$\Omega = \dfrac{d\theta}{dt} \quad$ angular velocity of precession

$\Omega = \dfrac{dL}{L \sin\phi} \dfrac{1}{dt} = \dfrac{\tau}{L \sin\phi} \quad$ torque provided by weight

$\Omega = \dfrac{r M g \sin\phi}{L \sin\phi}$

$\boxed{\Omega = \dfrac{M g r}{L}} \quad$ inversely proportional to angular momentum

$\sin(\pi - \phi) = \sin\phi$

Faster the top spins
→ greater the L
→ slower the precession

Block Inside a Hoop on top of a Table

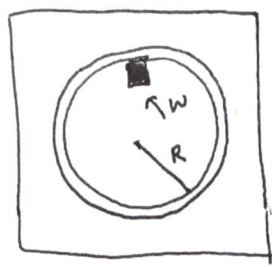

hoop - mass m
block - negligible mass

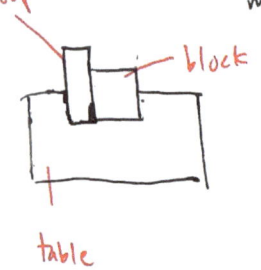

A hoop with Radius R is attached onto the top of a table. Inside the hoop, a block is placed against the side of the hoop. There is μ_k of kinetic friction between block, table, and inside surface of hoop. The block has initial angular velocity $\omega = \sqrt{\frac{g}{R}}$

Find $\omega(t)$, how angular velocity varies with time.

$N_{hoop} = mg$

$N_{hoop} = m\omega^2 R$

$\boxed{F_{fr\,hoop} = \mu m \omega^2 R}$

$N_{table} = mg$

$\boxed{F_{fr\,table} = \mu m g}$

$\tau = I\alpha$

$-R F_{fr\,hoop} - R F_{fr\,table} = mr^2 \alpha$

$-\mu m (R^2 \omega^2 + gR) = mr^2 \alpha$

$\alpha = -\mu(\omega^2 + \frac{g}{R})$

$\boxed{\frac{d\omega}{dt} = -\mu(\omega^2 + \frac{g}{R})}$

rearrange: $\dfrac{d\omega}{\omega^2 + \frac{g}{R}} = -\mu \, dt$

integrate: $\displaystyle\int_{\omega_i}^{\omega} \frac{d\omega}{\omega^2 + \frac{g}{R}} = \int_0^t -\mu \, dt$

$\sqrt{\frac{R}{g}} \left(\tan^{-1}(\omega \sqrt{\frac{R}{g}}) - \overbrace{\tan^{-1}(\omega_0 \sqrt{\frac{R}{g}})}^{\tan^{-1}(1)} \right) = -\mu t$

$\sqrt{\frac{R}{g}} \left(\tan^{-1}(\omega \sqrt{\frac{R}{g}}) - \frac{\pi}{4} \right) = -\mu t$

$\tan^{-1}(\omega \sqrt{\frac{R}{g}}) = -\mu t \sqrt{\frac{g}{R}} + \frac{\pi}{4}$

$\omega \sqrt{\frac{R}{g}} = \tan(\frac{\pi}{4} - \mu t \sqrt{\frac{g}{R}})$

$\boxed{\omega = \sqrt{\frac{g}{R}} \tan(\frac{\pi}{4} - \mu t \sqrt{\frac{g}{R}})}$

Total Time for block to stop.

$\omega(T) = 0$

$0 = \frac{\pi}{4} - \mu t \sqrt{\frac{g}{R}}$

$\boxed{T = \frac{\pi}{4\mu} \sqrt{\frac{R}{g}}}$

Solid Cylinder with 2 cords

solid cylinder: Length L
Radius R
Mass M

2 cords are wrapped around the cylinder - one attached at each end

a. Derive Moment of Inertia for Cylinder

$$I = \int r^2 \, dm$$

$$I = \int_0^R r^2 \frac{2Mr}{R^2} \, dr$$

$$I = \frac{2M}{R^2} \int_0^R r^3 \, dr$$

$$\boxed{I = \tfrac{1}{2} MR^2}$$

$\rho = \frac{m}{V}$

$m = \rho V$

$m = \rho \pi r^2 L$

$dm = \rho 2\pi r L$

$dm = \frac{2Mr \, dr}{R^2}$

ρ = mass density

$\rho = \frac{M}{\pi R^2 L}$

note: big R used bc specific to this cylinder

b. Find Tension in the cords as they unwind without slipping

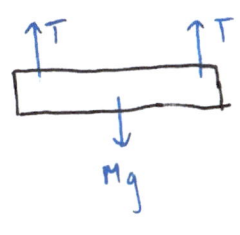

$\Sigma F:$ $Mg - 2T = Ma$

$a = g - \frac{2T}{M}$

$\Sigma \tau:$ $2TR = I\alpha$

$2TR = I \frac{a}{R}$ ← rolling w/o slipping

$2TR^2 = I \left(g - \frac{2T}{M} \right)$

$T = \frac{IMg}{2R^2 M + 2I}$ ← $I = \tfrac{1}{2} MR^2$

$$\boxed{T = \tfrac{1}{6} Mg}$$

c. Find linear acceleration as cylinder falls

$$a = g - \frac{2T}{M} = g - \frac{2}{M}\left(\tfrac{1}{6} Mg\right)$$

$$\boxed{a = \tfrac{2}{3} g}$$

*linear acceleration is a, not α

Putty attaches itself at rod's CM. How high does end of rod swing?

rod length = l *similar to ballistic pendulum

m, v

angular momentum conserved

$L_{before} = L_{after}$

$m(\frac{1}{2}l)v = (I_{rod} + I_{putty})\omega$

$$\boxed{\omega = \frac{mlv}{2(I_{rod} + I_{putty})}}$$

energy conservation

$KE_{after\ collision} = PE_{top\ of\ swing}$

$\frac{1}{2}(I_{rod} + I_{putty})\omega^2 = (m+M)gh_{CM}$

rotational KE ←   → gravitational PE

$$h_{CM} = \frac{m^2 v^2}{2g(m+M)(\frac{4}{3}M+m)}$$

$$\boxed{h_{bottom} = 2h_{CM}}$$

Ballistic Pendulum

A bullet of mass m buries itself in a block of wood of mass M. The system $(m+M)$ swings up.

linear momentum conservation

$mv_0 = (m+M)v_f$

$v_f = \dfrac{mv_0}{m+M}$

Note: All KE of system is transformed into gravitational PE.

energy conservation

$\frac{1}{2}(m+M)v_f^2 = (m+M)gh$

$v_f = \sqrt{2gh}$

equate equations

$\dfrac{mv_0}{m+M} = \sqrt{2gh}$

$$\boxed{v_0 = \frac{(m+M)\sqrt{2gh}}{m}}$$ velocity of bullet

Fluids, Oscillations, and Bernoulli's

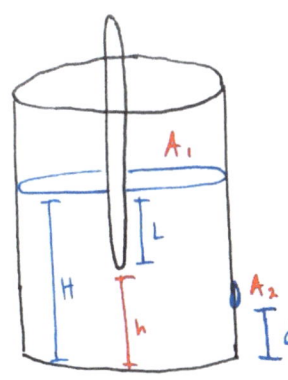

log is placed into fluid of density ρ_L
h = distance from log to bottom of bucket
L = submerged portion
H = height of fluid in bucket

a. If the log is <u>displaced upward</u> from equilibrium by <u>distance x</u>, find net force on log.

Key: F_B will decrease by: $\Delta F_{buoyant} = (A_{straw} \, x) \rho_L g$

next step: find A_{straw}

at equilibrium: $m_L g = L A_{straw} \rho_L g$ (weight of straw)

$A_{straw} = \dfrac{m_L}{L \rho_L}$

substitute → $\Delta F = \dfrac{-m_L \rho_L g}{L \rho_L} x$

$$\boxed{\Delta F = \dfrac{-m_L g}{L} x}$$

negative ← restoring force

b. Is log SHO?

$F = ma$ → $\dfrac{-m_L g}{L} x = ma$ → $\boxed{\dfrac{d^2x}{dt^2} + \dfrac{g}{L} x = 0}$ matches differential equation of SHO!

$\omega = \sqrt{\dfrac{g}{L}}$ → $T = 2\pi \sqrt{\dfrac{L}{g}}$

c. Drill a hole into side of Bucket at height d of area A_2

$\cancel{P_{atm}} + \rho_L g H + \frac{1}{2} \rho v_{top}^2 = \cancel{P_{atm}} + \rho_L g d + \frac{1}{2} \rho v_{hole}^2$

if $A_2 \ll A_1$: $\rho_L g (H-d) = \frac{1}{2} \rho_L v_{hole}^2$

$\boxed{v_{hole}^2 = 2g(H-d)}$

Range → use 1D kinematics

$t = \sqrt{\dfrac{2d}{g}}$ $\boxed{R = 2\sqrt{d(H-d)}}$

if $A_2 v_{hole} = A_1 v_{top}$

$\rho g (H-d) = \frac{1}{2} \rho_L \left[v_{hole}^2 - \left(\dfrac{A_2}{A_1}\right)^2 v_{hole}^2 \right]$

$\boxed{v_{hole}^2 = \dfrac{2g(H-d)}{1 - \left(\dfrac{A_2}{A_1}\right)^2}}$

Elevator Fluids

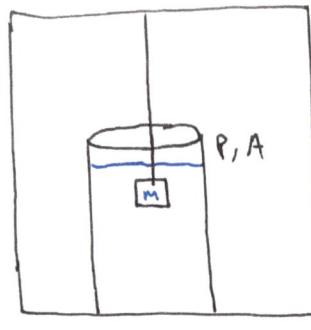

block of mass m is immersed in a fluid in a bucket in an elevator

a. $F_B + T = mg \qquad F_B = mg = \rho V g$

$\rho V g + T = mg$

$\boxed{V = \dfrac{mg - T}{\rho g}}$ volume of block

b. Find Tension in cord if elevator is accelerating upward with $a = \tfrac{1}{3}g$

$F_B + T_2 - mg = m\left(\tfrac{g}{3}\right)$ note: $T_2 \neq T$ as in a.

$T_2 = \tfrac{4}{3}mg - \rho V g$

$\boxed{T_2 = T + \tfrac{1}{3}mg}$

c. now elevator is accelerating downward at $a = \tfrac{1}{3}g$

$F_B + T_3 - mg = m\left(\tfrac{-g}{3}\right)$

$\boxed{T_3 = T - \tfrac{1}{3}mg}$

d. elevator in free fall?

$\boxed{T_4 = 0}$

e. Find the time it takes for the beaker to empty

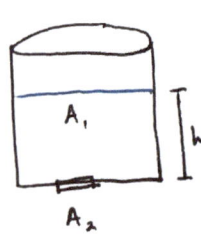

1. $\rho g h + \tfrac{1}{2}\rho v_1^2 = \tfrac{1}{2}\rho v_2^2$

   $2gh = v_2^2 - v_1^2$

2. $A_1 v_1 = A_2 v_2$

   $v_2 = \dfrac{A_1}{A_2} v_1$

3. $v_1 = \dfrac{-dh}{dt}$

   note: Neg. sign

Plug ③ → ①

$2gh = v_1^2 \left(\dfrac{A_1^2}{A_2^2} - 1\right)$

③ → $\dfrac{-dh}{dt} = \sqrt{\dfrac{2g A_2^2 h}{A_1^2 - A_2^2}} \qquad k = \sqrt{\dfrac{2g A_2^2}{A_1^2 - A_2^2}}$

$\dfrac{-dh}{\sqrt{h}} = k\, dt$

$-\int_{h_i}^{0} \dfrac{dh}{\sqrt{h}} = \int_0^{t_f} k\, dt$

$\boxed{t_f = \sqrt{\dfrac{2h}{g}\left(\dfrac{A_1^2 - A_2^2}{A_2^2}\right)}}$

Plank on Bicycle Wheels

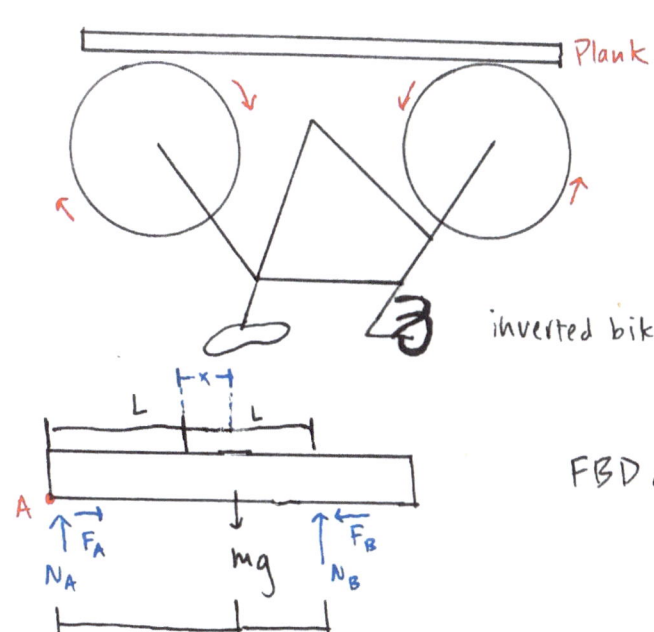

inverted bike

FBD of plank

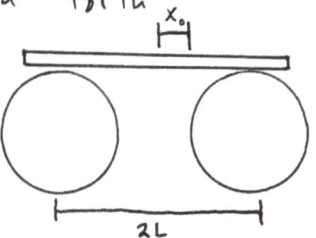

A Plank of mass m rests on 2 identical bicycle wheels that turn rapidly in opposite directions. The plank oscillates back and forth.

Initially the plank is held at rest w/ its center at distance x_0 from midpoint of the wheels

a. Find normal force on each wheel

$\Sigma \tau$: $mg(L+x) = N_B(2L)$ pivot about pt A

$$\boxed{N_B = \frac{gm}{2}\left(1 + \frac{x}{L}\right)}$$

ΣF: $N_A + N_B = mg$

$N_A = mg - \frac{gm}{2}\left(1 + \frac{x}{L}\right)$

$$\boxed{N_A = \frac{mg}{2}\left(1 - \frac{x}{L}\right)}$$

b. Find Frictional forces

$F_A = \mu N_A$

$$\boxed{F_A = \frac{\mu mg}{2}\left(1 - \frac{x}{L}\right)}$$

$F_B = \mu N_B$

$$\boxed{F_B = \frac{\mu gm}{2}\left(1 + \frac{x}{L}\right)}$$

c. Does the plank show simple harmonic oscillation?

$\Sigma F = ma = -F_B + F_A$

$m\frac{d^2x}{dt^2} = \frac{mg\mu}{2}\left[\left(1 - \frac{x}{L}\right) - \left(1 + \frac{x}{L}\right)\right]$

rearrange:

$\frac{d^2x}{dt^2} = \frac{-g\mu}{2} \cdot \frac{2x}{L}$

$$\boxed{\frac{d^2x}{dt^2} + \frac{\mu g}{L}x = 0}$$ ✓ matches diff. eq of SHO

$\omega = \sqrt{\frac{\mu g}{L}}$

$T = 2\pi\sqrt{\frac{L}{\mu g}}$

d. Find amplitude and phase angle (ϕ)

initial conditions: at $t=0$ $x = x_0$
$\frac{dx}{dt} = 0$

$x = A\cos(\omega t + \phi)$ initial conditions $t=0$ $x_0 = A\cos(\phi)$ 1

$\frac{dx}{dt} = -A\omega \sin(\omega t + \phi)$ $0 = -A\omega \sin(\phi)$ 2

to satisfy 1 & 2

$$\boxed{x(t) = x_0 \cos\left(\sqrt{\frac{\mu g}{L}}\, t\right)}$$

$\phi = 0$
$A = x_0$

Spring Attached to a Spring

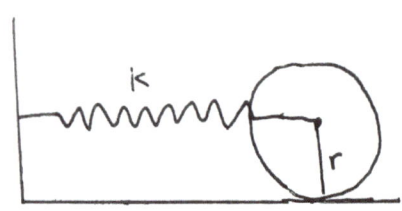

A sphere of mass M is attached to a spring of constant k. The spring is attached to the sphere at its center. There is μ_s between table and sphere. Find maximum speed of sphere if it rolls without slipping.

rolling without slipping condition:
$v = r\omega$ and $a = r\alpha$

Key: Relate Torque and Force

$\tau = I\alpha$ $\qquad F = ma$ $\quad$ equate equations: $\quad \dfrac{\tau r}{I} = \dfrac{F}{m}$

$\tau = I\dfrac{a}{R}$ $\qquad a = \dfrac{F}{m}$

$a = \dfrac{\tau r}{I}$ $\qquad\qquad\qquad\qquad \dfrac{r}{\frac{2}{5}mr^2}\tau = \dfrac{1}{m}F$

$\qquad\qquad\qquad\qquad\qquad\qquad \dfrac{5}{2mr}\Sigma\tau = \dfrac{1}{m}\Sigma F$

$F_{spring} = -kx$

$-kx = -\dfrac{7}{2}\mu_s mg$ $\qquad\qquad \dfrac{5}{2mr}(-rf_{sf}) = \dfrac{1}{m}(f_{sf} - F_{spring})$

$\boxed{x = \dfrac{7\mu_s mg}{2k}}$ $\qquad\qquad$ rearrange: $\quad \boxed{F_{spring} = \dfrac{7}{2}f_{sf}}$

Spring PE when stretched distance x:
$E = \tfrac{1}{2}kx^2 = \tfrac{1}{2}k\left(\dfrac{7\mu_s mg}{2k}\right)^2$

$\boxed{E = \dfrac{49\mu^2 m^2 g^2}{8k}}$

Energy when sphere passes equilibrium pt: ALL KE from translation and rotation.

$E = \tfrac{1}{2}mv_{max}^2 + \tfrac{1}{2}I\omega^2$

$= \tfrac{1}{2}mv_{max}^2 + \tfrac{1}{2}\cdot\tfrac{2}{5}mr^2\left(\dfrac{v_{max}}{r}\right)^2$

$\boxed{E = \dfrac{7}{10}mv_{max}^2}$

Energy conservation: $\quad \dfrac{7}{10}mv_{max}^2 = \dfrac{49\mu^2 m^2 g^2}{8k} \rightarrow \boxed{V_{max} = \sqrt{\dfrac{35\mu^2 mg^2}{8k}}}$

Oscillations of a Rod attached to a Spring

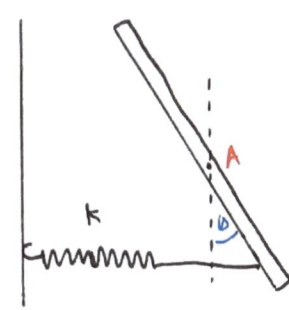

Rod - Mass M, length l

no friction

assume $\sin\theta \approx \theta$ and $\cos\theta \approx 1$ by small angle approximation

a. Find angular frequency (ω) of rod's oscillations in terms of parameters given

define x: as extension of the spring

$x = \frac{l}{2}\sin\theta$

$x \approx \frac{l}{2}\theta$

$\tau = kx \cdot \frac{l}{2}\cos\theta$ — taken about pt A — caused by spring

$\tau \approx \frac{kl^2\theta}{4}$

$\tau = I\alpha$

$\frac{kl^2\theta}{4} = -\frac{1}{12}ml^2\alpha$ — negative bc it's a restoring force

$\alpha = -\frac{3k}{m}\theta$ — differential equation matching that of a SHO!

$\alpha = -\omega^2\theta$

$$\boxed{\omega = \sqrt{\frac{3k}{m}}}$$

b. scenario 2: cut off top half of rod because you feel like it. Find angular frequency of remaining half of rod.

Key: gravity acting on CM now causes a torque

$M \to \frac{m}{2}$ $\quad l \to \frac{l}{2}$

CM = $\frac{l}{4}$ from hinge

moment of Inertia: $I = \frac{1}{3}\cdot\frac{m}{2}\cdot\frac{l^2}{2^2}$

$I = \frac{1}{24}ml^2$

$\tau = \frac{m}{2}g\cdot\frac{l}{4}\sin\theta$

$\tau = \frac{mgl}{8}\theta$ — caused by weight (not present in part 1)

total torque: $\frac{mgl}{8}\theta + \frac{kl^2}{4}\theta$

$\tau = I\alpha$

$\tau = \frac{1}{24}ml^2\alpha$

$\frac{mgl}{8}\theta + \frac{kl^2}{4}\theta = -\frac{1}{24}ml^2\alpha$

$\alpha = -\frac{\left(\frac{mgl}{8}+\frac{kl^2}{4}\right)}{\frac{1}{24}ml^2}\theta \Rightarrow \alpha = -\omega^2\theta$

$$\boxed{\omega = \sqrt{\frac{6k}{m} + \frac{3g}{l}}}$$

The Torsional Pendulum

Disk is suspended from a wire and is allowed to twist.

given: $\tau = -k\theta$

$\tau = I\alpha$

$I \dfrac{d^2\theta}{dt^2} = -k\theta$

$$\boxed{I \dfrac{d^2\theta}{dt^2} + k\theta = 0}$$ ✓ form of SHO differential eq.

$$\boxed{\omega^2 = \dfrac{k}{I}}$$

Vertical Spring + Horizontal Rod

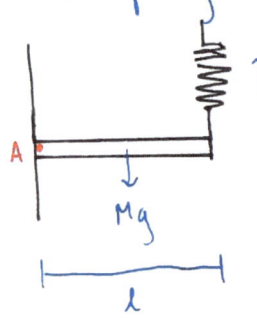

Is this system a SHO? If so, find diff eq. and frequency.

Key: find torque about pt. A

y_0: amt spring is stretched at equilibrium

y: additional distance spring is stretched below horizontal

$\tau = I\alpha$

$\tfrac{1}{2}\ell Mg - k(y+y_0)\ell = I\alpha$

$\tfrac{1}{2}\ell Mg - k(y+y_0)\ell = \tfrac{1}{3}M\ell^2 \dfrac{d^2\theta}{dt^2}$

at equilibrium:

$\Sigma\tau: Mg(\tfrac{1}{2}\ell) = ky_0\ell$

$\tfrac{1}{2}\ell Mg - ky\ell - ky_0\ell = \tfrac{1}{3}M\ell^2 \dfrac{d^2\theta}{dt^2}$

$\tfrac{1}{2}\ell Mg - ky\ell - \tfrac{1}{2}\ell Mg = \tfrac{1}{3}M\ell^2 \dfrac{d^2\theta}{dt^2}$

$-ky\ell = \tfrac{1}{3}M\ell^2 \dfrac{d^2\theta}{dt^2}$

✓ form of SHO diff. eq.

$$\boxed{\dfrac{d^2\theta}{dt^2} + \dfrac{3k}{M}\theta = 0}$$

$$\boxed{\omega = \sqrt{\dfrac{3k}{M}}}$$

$$\boxed{f = \dfrac{1}{2\pi}\sqrt{\dfrac{3k}{M}}}$$

Collisions and Oscillations

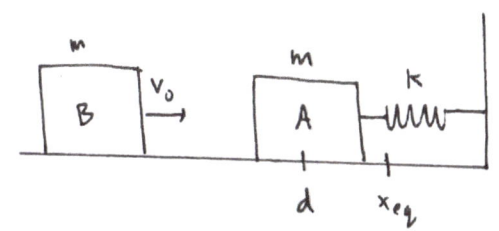

frictionless surface

mass m (A) is displaced d from equilibrium and released to oscillate

Simple Harmonic Motion present: Key

Amplitude: d $\qquad w = \sqrt{\frac{k}{m}}$

$$\boxed{\begin{array}{l} x(t) = d\cos(wt) \\ v(t) = -wd\sin(wt) \end{array}} \qquad T = 2\pi\sqrt{\frac{m}{k}}$$

Special Scenario: Friction is Present

$F = -bv$: damping force

$-kx - bv = ma$

$$m\frac{d^2x}{dt^2} + b\frac{dx}{dt} + kx = 0$$

differential equation!

$$\boxed{w' = \sqrt{\frac{k}{m} - \frac{b^2}{4m^2}}} \quad \text{new angular frequency}$$

Collision: When Block A is at furthest left displacement
: Inelastic

lin. momentum conserved $\qquad mv_0 = 2mv_1$

$\qquad v_1 = \frac{v_0}{2}$ to the right

energy conserved $\qquad \frac{1}{2}kd^2 + \frac{1}{2}(2m)v_1^2 = \frac{1}{2}kA^2$

spring PE + translational KE of masses = max compression of spring PE

$$\boxed{A = \sqrt{d^2 + \frac{1}{2}\frac{m}{k}v_0^2}}$$

$$w = \sqrt{\frac{k}{2m}}$$

$$T = 2\pi\sqrt{\frac{2m}{k}} \qquad T = \frac{2\pi}{w}$$

Collision: When Block A is passing equilibrium position to the right

key: At moment of collision, block A will have v_{max}, all energy will be KE

$v_{max} = Aw = dw = d\sqrt{\frac{k}{m}}$

lin. momentum conserved $\qquad mv_0 + md\sqrt{\frac{k}{m}} = 2mv_1$

$$\boxed{v_1 = \frac{1}{2}\left(v_0 + d\sqrt{\frac{k}{m}}\right)} \text{ to the right}$$

* This problem is key to understanding the properties of a mass along its positions of oscillation and how an adding mass will influence them

Ball in a Bowl

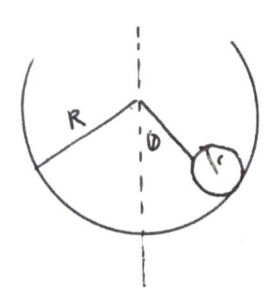

Ball: radius r, mass m placed in bowl of radius R
μ_s is present between ball and bowl

a. Is there simple harmonic motion of the ball? If so, derive an equation.

Key: find differential equation that matches that of a SHO
trick: use energy

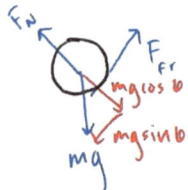

$h = R - R\cos\theta$

$$U = mgh = mgR(1-\cos\theta) = mgR\frac{\theta^2}{2}$$

$$KE = \tfrac{1}{2}mv^2 + \tfrac{1}{2}I\omega^2 = \tfrac{1}{2}mv^2 + \tfrac{1}{2}\cdot\tfrac{2}{5}mr^2\omega^2 = \tfrac{1}{2}\cdot\tfrac{7}{5}mR^2\omega^2$$

$E = K + U$ conservation of energy

$$\frac{d}{dt}\left(mgR\frac{\theta^2}{2} + \tfrac{1}{2}\cdot\tfrac{7}{5}mR^2\omega^2\right) = 0$$

take derivative w/ respect to time — apply chain Rule

$$mgR\theta\omega + \tfrac{7}{5}mR^2\omega\alpha = 0$$

$$g\theta + \tfrac{7}{5}R\alpha = 0$$

✓ matches diff. equ. of SHO!

$$\boxed{\alpha + \tfrac{5}{7}\tfrac{g}{R}\theta = 0} \rightarrow \omega^2 = \tfrac{5}{7}\tfrac{g}{R}$$

$$\boxed{\omega = \sqrt{\tfrac{5}{7}\tfrac{g}{R}}}$$

b. find max θ_0 such that ball will <u>start slipping</u>

ΣF X: $mg\sin\theta - F_{fr} = ma$
   Y: $N = mg\cos\theta$

$\Sigma \tau$: $\tfrac{2}{5}mR^2\alpha = F_{fr}R$
   $a = \alpha R$
   $\rightarrow a = \tfrac{5}{2}\tfrac{F_{fr}}{m}$

Key: when ball is about to slip, F_{fr} will become max. value

$$mg\sin\theta = \tfrac{7}{2}F_{fr}$$

$$mg\sin\theta_0 = \tfrac{7}{2}\mu_s mg\cos\theta_0$$

small angle approximation
$\sin\theta \sim \theta$
$\cos\theta \sim 1$

$$\boxed{\theta_0 = \tfrac{7}{2}\mu_s}$$

alternate solution to part a. → using torque strategy rather than energy

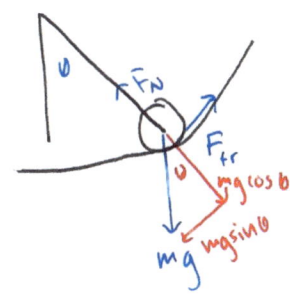

x: $mg\sin\theta - F_{fr} = -ma$

y: $N = mg\cos\theta$

τ: $I\alpha = F_{fr} r$

$\frac{2}{5} mr^2 \alpha = F_{fr} r$

$-a = \alpha r$

→ $F_{fr} = -\frac{2}{5} ma$ → plug into x equation

torque is acting ball of radius r

$\theta + \frac{7}{5} \frac{R}{g} \alpha = 0$

$\boxed{\alpha + \frac{5}{7} \frac{g}{R} \theta = 0 \qquad w = \sqrt{\frac{5}{7} \frac{g}{R}}}$

Get rid of The Cockroach

There is a cockroach standing on a tightrope stretched with tension T_s. The tightroper moves her foot up and down creating a sinusoidal transverse wave (λ, A). Find the minimum wave amplitude (A_{min}) such that the cockroach will become momentarily weightless as the wave passes underneath it.

Know: $v = \sqrt{\dfrac{T_s}{\mu}}$ $\omega = 2\pi \dfrac{v}{\lambda}$ $a = \omega^2 A_{min}$

property of transverse waves

second derivative of $x(t)$ wave function

Key: at that moment when cockroach becomes momentarily weightless, the string is accelerating more than gravity

$$g = \omega^2 A_{min}$$

$$A_{min} = \dfrac{g}{\omega^2} = \dfrac{g\lambda^2}{4\pi^2 v^2} \longrightarrow \boxed{A_{min} = \dfrac{g\lambda^2 \mu}{4\pi^2 T_s}}$$

2 Springs Attached to a Block

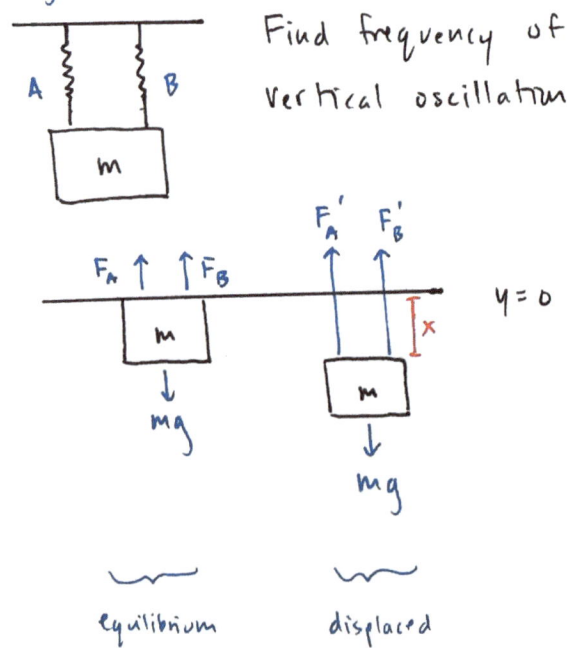

equilibrium   displaced

Find frequency of vertical oscillation.

at equilibrium: $F_A + F_B = mg$

displaced x from equilibrium:

$F_A' + F_B' = mg$

$F_A - kx + F_B - kx = mg$

$-2kx + F_A + F_B = mg$

This is a version of SHM w/ spring constant of $2k$

$$\boxed{f = \dfrac{1}{2\pi}\sqrt{\dfrac{2k}{m}}}$$

Interaction of Reflected and Transmitted Waves

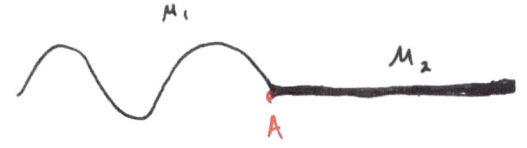

A cord is made of 2 sections of 2 linear densities: μ_1 and μ_2. An oscillator creates a sinusoidal wave on the left of the cord that travels right. When it reaches pt A part of the wave is reflected and part of it is transmitted into section (μ_2).

Wave equation for reflected wave

$$x(t) = A_R \sin(k_1 x + \omega t)$$

key: left traveling – note sign of ω

Wave equation for transmitted

$$x(t) = A_T \sin(k_2 x - \omega t)$$

key: right traveling – note sign of ω

$$y_R(x,t) = A_T \sin(k_2 x - \omega t)$$

Superposition of Incident and Reflected wave

$$y_L(x,t) = A \sin(k_1 x - \omega t) + A_R \sin(k_1 x + \omega t)$$

Find A_T and A_R

initial conditions for the wave equation:

at $x = 0$: $\quad y_L(0,t) = y_R(0,t) \quad$ — initial displacement

$$(A\sin(-\omega t) + A_R \sin(\omega t) = A_T \sin(-\omega t)) / \sin(\omega t)$$

① $\quad -A + A_R = -A_T$

remember: $\sin(\omega t)$ is odd
$\sin(-\omega t) = -\sin(\omega t)$

at $x = 0$: $\quad \dfrac{dy_L}{dx}(0,t) = \dfrac{dy_R}{dx}(0,t) \quad$ — initial velocity

remember
$\cos(\omega t)$ is even
$\cos(-\omega t) = \cos(\omega t)$

$$(k_1 A \cos(-\omega t) + k_1 A_R \cos(\omega t) = k_2 A_T \cos(-\omega t)) / \cos(\omega t)$$

② $\quad k_1(A + A_R) = k_2 A_T$

① & ② →

$$\boxed{A_R = \dfrac{1 - \dfrac{k_1}{k_2}}{1 + \dfrac{k_1}{k_2}} A}$$

$$\boxed{A_T = \dfrac{2 \dfrac{k_1}{k_2}}{1 + \dfrac{k_1}{k_2}} A}$$

Energy of a small segment of traveling wave in a string

key: Energy associated with a traveling wave is a sum of both Potential and Kinetic Energy

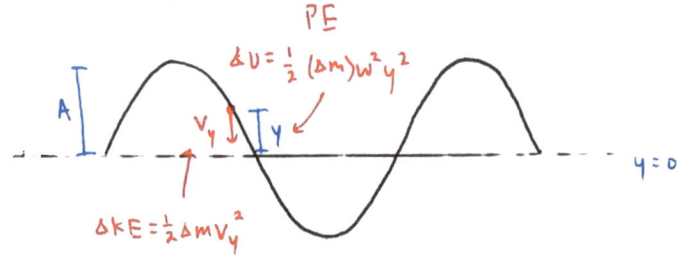

$dU = \frac{1}{2}(\Delta m)\omega^2 y^2$ — PE

$\Delta KE = \frac{1}{2}\Delta m v_y^2$

differential form of PE:

$dU = \frac{1}{2}\omega^2 y^2 \, dm$

$dU = \frac{1}{2}\omega^2 y^2 \mu \, dx$

form of traveling wave
$y = A\sin(kx - \omega t)$

$dU = \frac{1}{2}\mu\omega^2 A^2 \sin^2(kx-\omega t)\,dx$

$U_\lambda = \frac{1}{2}\mu\omega^2 A^2 \int_0^\lambda \sin^2(kx)\,dx$

integrate to find energy for one full wavelength

$$\boxed{U_\lambda = \frac{1}{4}\mu\omega^2 A^2 \lambda}$$

differential form of KE:

$dKE = \frac{1}{2}v_y^2 \, dm$

$dKE = \frac{1}{2}v_y^2 \mu \, dx$

derivative of wave equation
$v_y = \omega A \cos(kx-\omega t)$

$dKE = \frac{1}{2}\mu\omega^2 A^2 \cos^2(kx-\omega t)\,dx$

$KE_\lambda = \frac{1}{2}\mu\omega^2 A^2 \int_0^\lambda \cos^2(kx)\,dx$

$$\boxed{KE_\lambda = \frac{1}{4}\mu\omega^2 A^2 \lambda}$$

integrate for energy of a full wavelength

$E_\lambda = U_\lambda + KE_\lambda$

Note: PE and KE are the same

$$\boxed{E_\lambda = \frac{1}{2}\mu\omega^2 A^2 \lambda}$$

The Rolling Sphere Paradox

We already know that a rolling sphere must eventually come to rest as it cannot roll forever. But is friction the force that causes the ball to slow down?

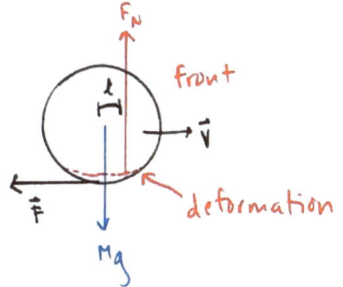

$\tau = Fr$

Paradox: $\vec{F}$ decelerates <u>translational</u> motion

but

$\vec{F}$ also accelerates <u>rotational</u> motion

Resolution: Must consider weight by gravity (mg) and normal force (F_N)

Mg & F_N: vertical forces and thus do not affect horizontal translational motion

: do NOT produce torque (no lever arm)

1. all objects are deformable
   Upon contact with the surface, the sphere will slightly flatten thus producing an <u>area, rather than point, of contact</u>. <u>F_N will act along this area</u>, producing a torque that slows down the sphere.

2. F_N will act in front of Mg
   Because the sphere is rolling to the right, the front or right side surface of the sphere will strike the table harder creating a small impulse. Thus, the <u>table will push upwards harder on the front of the ball</u>. This also produces a slight torque that slows down the ball.

Double Atwood Machine

frictionless, massless pulleys and cords

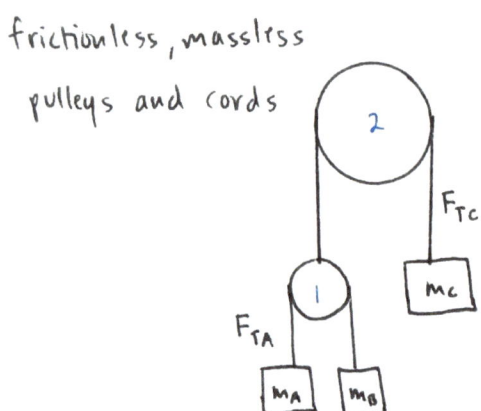

FBD's

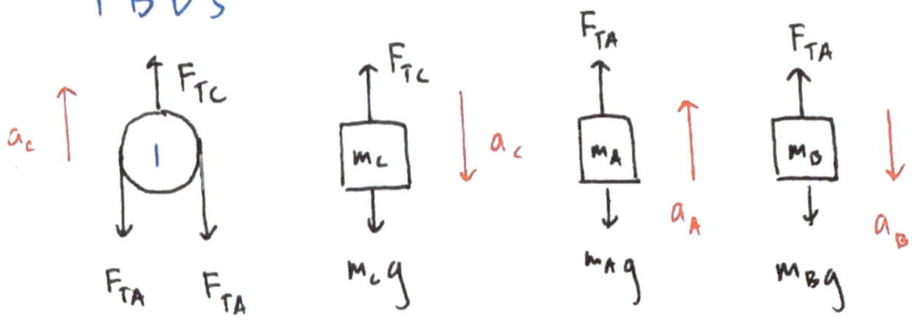

(direction of the object's acceleration is the positive direction of motion)

Key: direction of the object's acceleration is the **positive direction** of motion

m_A: $F_{TA} - m_A g = m_A a_A$

$F_{TA} - m_A g = m_A (a_R + a_C)$

m_B: $m_B g - F_{TA} = m_B a_B$

$m_B g - F_{TA} = m_B (a_R - a_C)$

m_C: $m_C g - F_{TC} = m_C a_C$

write as system of linear equations:

★ $a_A = a_R + a_C$

★ $a_B = a_R - a_C$

a_R = acceleration relative to the pulley

↳ negative bc a_C of pulley 1 moves in opposite direction of a_B

pulley: $2 F_{TA} = F_{TC}$

$$F_{TA} - m_A a_C - m_A a_R = m_A g$$
$$F_{TA} - m_B a_C + m_B a_R = m_B g$$
$$2 F_{TA} + m_C a_C \qquad\qquad = m_C g$$

Solve system using Cramer's Rule:

$$a_C = \frac{\begin{vmatrix} 1 & m_A & -m_A \\ 1 & m_B & m_B \\ 2 & m_C & 0 \end{vmatrix}}{\begin{vmatrix} 1 & -m_A & -m_A \\ 1 & -m_B & m_B \\ 2 & m_C & 0 \end{vmatrix}} = \frac{m_A m_C + m_B m_C - 4 m_A m_B}{4 m_A m_B + m_A m_C + m_B m_C} g$$

$$a_R = \frac{\begin{vmatrix} 1 & -m_A & m_A \\ 1 & -m_B & m_B \\ 2 & m_C & m_C \end{vmatrix}}{\begin{vmatrix} 1 & -m_A & -m_A \\ 1 & -m_B & m_B \\ 2 & m_C & 0 \end{vmatrix}} = \frac{2(m_A m_C - m_B m_C)}{4 m_A m_B + m_A m_C + m_B m_C} g$$

$$F_{TA} = \frac{\begin{vmatrix} m_A & -m_A & -m_A \\ m_B & -m_B & m_B \\ m_C & m_C & 0 \end{vmatrix}}{\begin{vmatrix} 1 & -m_A & -m_A \\ 1 & -m_B & m_B \\ 2 & m_C & 0 \end{vmatrix}} = \frac{4 m_A m_B m_C}{4 m_A m_B + m_A m_C + m_B m_C} g$$

$$\boxed{F_{TC} = 2 F_{TA}}$$

$a_A = a_R + a_C = \dfrac{3 m_A m_C - m_B m_C - 4 m_A m_B}{4 m_A m_B + m_A m_C + m_B m_C} g$

$a_B = a_R - a_C = \dfrac{m_A m_C - 3 m_B m_C + 4 m_A m_B}{4 m_A m_B + m_A m_C + m_B m_C} g$

www.ingramcontent.com/pod-product-compliance
Lightning Source LLC
Chambersburg PA
CBHW051019180526
45172CB00002B/404